PLANET ZIGNOX

SPECIAL DELIVERY
Putting Math to Work

BY BOB KRECH

HAUNTED MANSION

RECYCLING ROOM

Dedication

This book is dedicated with love and thanks to Karen, Andrew, and Faith, my partners in this and everything else.

Note

For information about workshops, resource materials, and other services, write or call Bob Krech, 16 Empress Lane, Lawrenceville, NJ 08648. Phone (609) 883-0637.

Managing Editor: Alan MacDonell
Development Editor: John Nelson
Acquisitions Manager: Doris Hirschhorn

Design Manager: Phyllis Aycock
Cover Design and Illustrations: Tracey Munz
Text Design and Production: Fiona Santoianni

Copyright © 1998 by
Cuisenaire Company of America
125 Greenbush Road South, Orangeburg, NY 10962

2 3 4 5 SG 02 01 00 99

Contents

Acknowledgments

Thanks to Dr. Nancy Katims and Dr. Michael Katims for the initial inspiration and continued interest and support. Thanks also to all the students, teachers, and administrators at Dutch Neck School and throughout the West Windsor-Plainsboro Schools who have helped in the development and implementation of these ideas. Particular thanks to Janet Jackson for funding problem development work with teachers.

The original idea for *Shapey* was developed by Bob Krech, Freddie DiGeronimo, Anna Marie Dunn, Janet Gill, Linda Nicholls, and Jim Occhino as a part of the New Jersey Mathematics Coalition Standards Dissemination Project.

The original idea for *A Very Special Zoo* was developed by Linda Masure and Wanda McGuinness, Dutch Neck School, West Windsor-Plainsboro Schools, Princeton Junction, NJ.

The original idea for *Recycling Robot* was developed by Bob Krech, Cynthia Smith, and Donna Terrible, West Windsor-Plainsboro Schools, Princeton Junction, NJ.

The original idea for *The Giant's Shoe* was developed by Terri McClendon and Judi Shilling, West Windsor-Plainsboro Schools, Princeton Junction, NJ.

The development and continued refining of *The Necklace Shop* included input from Christi Campbell, Rosemary Geel, Janet Jackson, Jung Lee, Susan Lemeshow, Maryann McMahon-Nestor, Ceil Questad, Laura Williams, and Wendy Wise, Dutch Neck School, West Windsor-Plainsboro Schools, Princeton Junction, NJ.

The idea for *Haunted Mansion* was developed by Felicia Lowenstein and Bob Krech.

An earlier version of *Planet Zignox* appeared in *Tomorrow's Lessons*, Annual Monograph, Association of Math Teachers of New Jersey, New Brunswick, NJ, 1997.

Thanks to my editors at Cuisenaire: Doris Hirschhorn, for getting the ball rolling and pushing it along; Alan MacDonell for great suggestions and—among other things—the title; and John Nelson for his word-by-word editing and week-by-week encouragement.

Chapter 1 - Introduction and Rationale

My Math Reaction

I have many fond memories of my own elementary school days. I remember all the important things, of course, like recess, kickball, chasing people, being chased, and what kind of lunch boxes I had. My former teachers may be somewhat satisfied to know that I also have some happy recollections about the academics.

For example, I remember how excited I felt when we used those shiny black and chrome, real metal microscopes in Science. I was sure that any minute I would discover a cure for something, even though I was in fifth grade and looking at creek water from behind the playground.

I also vividly remember how in fourth grade during the height of the James Bond and *Man From UNCLE* craze, I would fill my writing class notebooks with page after page of my own cleverly crafted spy stories. (Most ended with the hero escaping through a trap door!) I was quite sure I was destined to be the next Ian Fleming.

And I can still almost feel the shortness of breath I would get in sixth grade when reading Edgar Allen Poe stories like *The Tell-Tale Heart*, and scaring myself so badly that I would sleep "in a cautious manner" for weeks at a time.

There were lots of great experiences in school for me. Unfortunately, math was never one of them. My basic approach to math class was not unlike that of a soldier in combat: keep my head down, do what I have to do, and then get out of the way.

I learned math in such a dull and meaningless way that it eventually became difficult. As I progressed through my math classes each year this became more and more true. What we learned in math seemed unconnected to anything else in my life or in school. There was no interest or fun.

The double-whammy in all of this was that the math exercises I disliked most were those with the greatest potential for interest and application – "Word Problems." My view at the time was that, instead of just giving us the equations and letting us get on with trying to work them out (which was bad enough), the equations were now embedded in nonsensical but complicated scenarios. The question on most everyone's mind in these little stories seemed to be: "When would two particular trains, traveling at different speeds, pass each other?" I was not enthralled. Why couldn't math be as exciting as writing was? Why couldn't math be fun like reading? Why couldn't math be interesting like science? The answer, of course, was that it could and it can.

Because of my own experience, when I started teaching I had what I call my "Math Reaction." I wanted to make sure my students' math experiences were more meaningful than the ones I had. I wanted them to learn math well, while experiencing it in ways that were useful, fun, motivating, and still challenging. There are many successful means to this end, but Project Problem™ work has been my favorite.

What are Project Problems?

Just what are Project Problems? Project Problems are challenging, long-term (about five 40-minute periods), hands-on, activity-based, complex problems that provide students with an opportunity to apply their math knowledge in a situation where they will achieve a real outcome.

The context of a Project Problem is always one of natural interest to children, like space, haunted houses, jewelry, animals, and the like. Most importantly though, when children do the math, something happens as a result! A model is created! An animal is helped! A mystery is solved! A treasure is uncovered! When children do Project Problems they experience math as something that is fun and useful. Math becomes an approach, a way of thinking, and a tool they can use to make things happen.

For example, in this book you will find a Project Problem entitled *Planet Zignox*. In the problem, students receive a letter from friendly aliens. The aliens ask for help in creating their new school. Interesting creatures these aliens. The tallest is one centimeter high. They have a limited budget for construction of the school, and they want it built out of Cuisenaire® Rods. They promise to come back once the models are done and "richly reward" the students.

To successfully complete the *Planet Zignox* Project Problem, students have to use double-digit addition, work with perimeter and metric measurement, use repeated addition and perhaps multiplication, keep a running tally, and control both money amounts and measurement simultaneously. They work in groups for four or five days on this problem. Will they create models that meet the aliens' requirements and then be richly rewarded? Students will certainly try (and they'll care a lot more about it than about the time when the two trains will pass each other). That is a Project Problem.

Project Problems, the NCTM Standards, and the TIMSS Study

While Project Problems like *Planet Zignox* are fun, engaging, and challenging – and these are compelling reasons to use them – the NCTM Standards also provide strong support for this mode of instruction.

The National Council of Teachers of Mathematics published their *Curriculum and Evaluation Standards for School Mathematics* in 1989. This document has had a profound influence on instruction, teacher development, textbook and program creation, and just about every aspect of math education in this country.

Using Project Problems is one means of implementing many of the NCTM standards in a way that is fun and exciting. As mentioned earlier, one of the key aspects of Project Problems is that they provide students an opportunity to apply their math learning in a real way. In the NCTM *Curriculum and Evaluation Standards for School Mathematics*, there is a discussion early on about the basic assumptions that governed the selection and shaping of the standards. One of these assumptions is that "the K-4 curriculum should emphasize the application of mathematics."[1] It goes on to say,

> If children are to view mathematics as a practical, useful subject, they must understand that it can be applied to a wide variety of real world problems and phenomena. By applying mathematics they learn to appreciate the power of mathematics.[2]

This application of mathematics in Project Problems is in the context of solving a long-term, complex problem. This fits well with NCTM Standard Number One, "Mathematics as Problem Solving." Problem Solving is where the emphasis of the movement for math reform lies, so it is not by chance that it is the first standard listed. The standard reads:

In Grades K-4, the study of mathematics should emphasize problem solving so that students can

- use problem-solving approaches to investigate and understand mathematical content;
- formulate problems from everyday and mathematical situations;
- develop and apply strategies to solve a wide variety of problems;
- verify and interpret results with respect to the original problem;
- acquire confidence in using mathematics meaningfully.[3]

In addition, in the NCTM *Summary of Changes in Content and Emphasis in K-4 Mathematics*, there is a call for increased attention to:

Problem Solving

- Word problems with a variety of structures
- Use of everyday problems
- Applications
- Study of patterns and relationships
- Problem-solving strategies[4]

Project Problems provide a mode of instruction that readily addresses the NCTM Standards on problem solving. The nature and structure of this type of problem solving and the procedures outlined here for implementation also provide constant opportunities for application in significant ways of the other three process standards: Math as Communication, Math as Reasoning, and Mathematical Connections.

The structure and delivery of teaching with Project Problems differ from much of current U.S. teaching practice. *Pursuing Excellence* is a report of the initial findings from the 1996 Third International Mathematics and Science Study (TIMSS), the largest, most comprehensive, and most rigorous international study of its kind. One of the key findings of this important study relates directly to teaching math in a Project Problem manner.

Through many hours of videotaping U.S. classrooms, a typical sequence of instruction was observed. The teacher instructed students in a concept or skill, solved example problems with the class, and then assisted students as they practiced similar problems. In contrast, teachers in Japanese classrooms posed a complex, thought-provoking problem, students struggled with the problem, shared solutions, discussed them, and then summarized conclusions with the teacher.[5] The study comments, "Our (U.S.) mathematics teachers' typical goal is to teach students how to do something, rather than how to think about and understand mathematical concepts."[6]

With Project Problems (as with most Japanese classrooms observed in the TIMSS study), a complex problem is presented, and students are invited to solve the problem with whatever methods and strategies they devise. They then share their methods and solutions. This is a very different style of teaching which calls on a higher level of thinking and greater responsibility on the part of the student. While no one determining factor is responsible for the difference between scores of U.S. and Japanese students,[7] the very clear differences in instructional modes would argue at the very least that students must think more and work in a more challenging and creative way.

WHEN TO USE PROJECT PROBLEMS

You can use Project Problems at different times and for different instructional purposes. Three basic approaches that address different objectives are Introduction, Investigation, and Culmination/Assessment.

Introducing a Concept

Planet Zignox is a third-grade problem in this book that features perimeter as one of its key concepts. If students do not understand what perimeter is, they cannot solve the problem. Yet you can assign them this Project Problem. Why? The context of the problem (helping tiny aliens to create a model school) greatly motivates students. They *want to know* just what perimeter is and how to find it so they can solve the problem. In this case, the problem creates a need that serves as a spring-board to learning about and then practicing a new concept introduced in the context of the problem.

Investigating a Concept

An appropriate Project Problem can engage students in investigation and further exploration of a concept. The *Planet Zignox* problem described above involves using a great deal of double-digit addition to keep track of money and materials spent in construction of the model. Before beginning this problem, you can provide students with many double-digit addition experiences. With this background, students can then investigate how this concept can be applied in the context of the problem.

Culmination/Assessment

Although complex problems like Project Problems incorporate a variety of concepts, there is often one overriding math concept or theme such as geometry, measurement, or addition. Some teachers like to teach a unit on the concept and then use the Project Problem as a culminating assessment activity in which students apply what they have learned without teacher instruction or intervention.

You can bring together elements of each of these three approaches by using a Project Problem for multiple purposes. There are opportunities for discovery, investigation, application, and assessment in every Project Problem.

Chapter 1 Notes

[1]*Curriculum and Evaluation Standards for School Mathematics,* National Council of Teachers of Mathematics, Reston, VA 1989, p. 18

[2]*Curriculum and Evaluation Standards for School Mathematics,* p. 18

[3]*Curriculum and Evaluation Standards for School Mathematics,* p. 23

[4]*Curriculum and Evaluation Standards for School Mathematics,* p. 20

[5]*Pursuing Excellence,* U.S. Department of Education, National Center for Educational Statistics. U.S. Government Printing Office: Washington, DC, 1996, p. 42

[6]*Pursuing Excellence,* p. 72

[7]*Pursuing Excellence,* p. 68

Chapter 2 - Project Problem Procedures

This procedures outline can be a starting point for you to think about how to effectively structure the time spent on Project Problems. I have used this format with good success while remaining flexible as to the number of days and, sometimes, the order of activities.

I. Day 1 - Special Delivery

1. The Special Delivery

Most of us like to get mail. Children surely do. Each Project Problem is designed to be launched with the receipt of a "problem letter" addressed to the class. The problem is embedded in the context of the letter. In my own class, this is how we always receive our problems. Over the last few years, I have had many accomplices who have aided me in "getting" our problem letters as in the scenario below.

The intercom clicks on. The front office secretary's voice comes through loud and clear. The children stop and listen. This is the same person who calls out the bus numbers for dismissal and announces whether recess is inside or outside that day and other important things like that.

"Mr. Krech?"

"Yes?"

"There is a package here in the office for your class. Could you send someone down for it?"

"Sure."

Twenty-four hands are instantaneously in the air. I send three people. After all, who knows how big the package is?

Five minutes later the three happy chosen return. They carry a large manila envelope, covered with canceled stamps, stuffed to bursting, and addressed in thick, black marker letters. Where might it be from?

Elizabeth, who is always on top of things, says "I bet it's our newest Project Problem!" And she's right. We quickly form a circle, and then s-l-o-w-l-y, painstakingly open up the package and pass out the individual copies of the letter. We begin to read silently.

This is our introduction. Sometimes the problems arrive from the office, sometimes they're found sitting on our floor or outside our window. Sometimes they are in big envelopes, sometimes in boxes, sometimes even in little pie-plate space ships, but somehow they always . . . arrive! Package your classes' problem letters however you like. It's fun and a good way to start the problem off "with a bang."

A common question from students during the introduction of a problem is, "Is this real?" As you will see when you read the problems in this book, most of the scenarios described are very fanciful and imaginative. Space aliens request help with portraits and school designs. Haunted Mansion Fun House owners ask for assistance in planning a new attraction. How you handle this will depend on you, your students, their ages, and the way you work together.

My own experience with elementary school students convinces me that most of them have one foot in reality and the other somewhere else. I am no longer surprised by fourth grade boys almost as tall as me, playing with little match box cars and stuffed animals in their desks. They are still children, and imaginative play is still a big part of their lives. The Project Problem scenarios capitalize on this.

When entering the context of a problem, I have observed that most children will "play along." If asked, "Is this real?" I always explain that "No, it is not real, but it is fun to pretend," or "Do you think aliens are really coming down to see us on Friday?" (The usual answer is "No" or "I don't know." My usual reply, "Well then, we'll see.") That is usually enough. I should also mention here that these problems are always structured so there is closure or a "payoff" at the end. If the aliens say they are going to "richly reward" students, then they surely will, as you will see in the problems themselves.

You may be asking yourself, "Is a problem about helping aliens real-life problem solving?" Certainly aliens are not "really" asking the class to help them out in real life (at least not to my knowledge) as they are in some of the Project Problems in this book. However, the skills and concepts required and the actual problem situations (such as designing a building, creating a pattern, keeping track of money and objects at the same time) are very real. These very real situations and concepts are embedded in a scenario or story that is not real but is fun and interesting. The skills and structures requiring problem solving are realistic while the contexts are often more unreal.

2. Reading the Letter

Most math we encounter in the real world is embedded in a context. Usually this context is a written document such as a pamphlet, statement, bill, brochure, or advertisement. So it is with Project Problems; the document is the "problem letter."

I begin by giving the whole class an opportunity to silently read through the problem letter. In kindergarten or first grade where there are more non-readers, I read the letter aloud or read from a larger version printed on a chart. In upper grades, after a period of silent reading, students take turns reading the letter aloud. I listen carefully as students read and help out where necessary.

3. Discussing the Letter

We stop and discuss words we may not be familiar with, keeping a list of vocabulary on the board or in notebooks.

The Concept of the Client

In the adult world, most of us work to try to meet the needs of a client. This is a very natural situation. Even as teachers who are not involved in private enterprise, we know we have several clients: students, parents, and the community.

Most of the problems in this book unfold from a situation in which a "client" wants the students to do something. Clients in Project Problems are space aliens, owners of fictitious businesses, animals, and the like. The idea of a client provides a focus for the work, an origination point, and a reason why the work needs to be done.

4. So, What's the Problem?

Once everyone has finished reading the letter, it is time for some key questions and discussion. I usually begin by asking a very general, open-ended question like, "So, what's going on here?" or "What is this all about?" During this discussion, I help students to think about the specific issues of the problem by asking questions like these.

What are you being asked to do?

Is there an answer or solution you are supposed to find?

Is there an end product you are supposed to create?

Are there any specifics about the product? Size? Other parameters?

Who wants the solution or product?

How will you know if you have successfully completed the problem?

These are general examples of the type of questions used during the discussion. There are problem-specific questions included for each Project Problem.

5. What materials do you think you will need to solve the problem?

After I am reasonably sure that the class knows what they are being asked to do in the problem, I ask, "What do you need to do this problem? What are the tools and materials you have to have? Let me know what they are and why you need them. I'll write them down on the board, and I'll have them for you tomorrow." This helps students to begin to think about how they will actually go about solving the problem and what specific strategies and tools they might employ.

I can usually anticipate what is actually needed in terms of tools and materials, but it is important for students to begin thinking about and naming tools and materials. It is also important for them to be able to justify why they will need a particular item. This discussion usually ends Day 1 and prepares us for Day 2, the first actual "work" day.

II. Day 2 - Let's Get To Work!

1. Materials and Re-Cap

I begin Day 2 by laying out the requested materials and tools at the front of the room. I review the main objectives of the problem: Who is the client? What does he or she want? What will the product be? What are the key parameters or limits? At the end of this discussion, I assign students to their work groups and distribute folders to each group so that they have a place to save their written work.

2. Grouping Considerations

The problems in this book can be done by individuals, partners, or small groups. Each Project Problem includes a suggested grouping size, but these are flexible. There are advantages and disadvantages to each arrangement. Most research in cooperative learning suggests that group sizes of between two and four work best.[1]

Social, academic, and behavioral considerations usually guide the creation of groups. Many teachers like to create groups that include high ability, average ability, and lower ability students together with the intent that greater learning, discussion, and interaction will occur. Research bears this out, I think, particularly in classes where cooperative roles are well defined and teachers are proactive in helping groups work well.[2]

This grouping model is not "the one way" however. At times, I have observed that groups of similar-ability students, particularly those having difficulty with math, do higher quality work than when in mixed-ability groups. Some students feel more comfortable, less intimidated, and more willing to participate when with peers of similar ability. As with most things, it depends on the individual students as to what arrangement works best for them.

3. Evaluation Pre-Set

If students are going to be evaluated on their Project Problem work, it is important to help them understand ahead of time just how this will be done. The beginning of this first work session is a good time to review your expectations. This may include sharing your scoring rubric, exemplars, grading policies, or other factors that will effect evaluation and assessment.

4. Time to Work - The Teacher's Role During Work Sessions

Once groups are set, the materials ready, the problem understood, and the expectations clear, it is finally time to work. Students working in their groups will be very involved. This is a perfect opportunity for you to move through the room to observe, listen, assist, assess, and record. As you observe, consider the following questions:

Is each student engaged?

What tools are students using?

What organizational strategies are working well for groups?

What problem-solving approaches and strategies are being utilized?

What are groups doing well?

What difficulties are groups having?

What kinds of mistakes are being made?

What math is being used?

Work sessions provide teachers with the opportunity to observe, take notes, and also guide as needed. During this first work session, I try to take as much of a "hands-off approach" as possible. It is very valuable for students to work on their own as much as possible and "bump into things" as they go. It is typical of students to make a few mistakes along the way, to self-correct, to persevere and eventually arrive at a solution. Sometimes your assistance will be greatly needed, sometimes not at all. Or they may not arrive at a solution. Real problem-solving is messy. If we as teachers carry students too much of the way, we remove the opportunity for them to experience the success that can result from hard work and the joy of discovery it can bring.

On the other hand, you don't want to see a student stuck, deep in the mud, frustrated to the point of tears. Here is an instance of where the *art* of teaching comes in, the decision when to intervene, when not to, how much to. It's based on your personal knowledge of your students. As a general rule in these problems, however, it is best to give students wait time.

III. Day 3 - Time to Share

1. Share Session

At the beginning of Day 3, we form a circle on the floor at the front of the room. Students sit with their fellow group members and bring along any work they have completed to that point. I begin by reviewing the task again with students. Then I ask for groups to share about the following points and we discuss each.

How did your group begin?

What worked well for your group?

What problems are you having?

What recommendations do you have for groups encountering problems?

During this discussion, I reinforce and help explain the ideas and approaches that are brought up. We talk about why different ideas may or may not be desirable in these instances. This sharing of ideas and problems can help students learn from their peers, adjust their work accordingly, and continue on their way to success.

After our share session, we get back to work. Again, I move about the room, observing, guiding, and questioning.

2. Mini-Lessons

Usually by this second work session, if there are significant difficulties and opportunities for teaching, they have begun to surface. Now is the time to stop and teach mini-lessons if appropriate.

Mini-lessons are short, to-the-point lessons on a particular skill, concept, or idea. They can be directed to one group, a group of individuals, or to the whole class. Sometimes I will stop everything and gather everybody together again at the front of the room and talk about a problem or idea I have observed. Then I do a quick lesson addressing it.

The element that makes these little lessons powerful is that, in the context of trying to solve the Project Problem, this information has clear and immediate application, relevance, and value to students. They can see this is helpful information. They want to know it, and they are highly motivated to learn it because they know it will help them achieve their mission.

After the presentation of the mini-lesson, I follow up on groups that I feel really needed that mini-lesson to ensure that the connection from the lesson to the problem is taking place.

At the end of each work session day, I review work folders to see if there is any problem evident in the work that needs addressing during subsequent share sessions or with an individual group.

IV. Day 4 - Still Working

1. Share Session

We begin again with a share session and follow the same format as before, discussing things that work well, problems, techniques, and suggestions. Some items discussed will be the same, but most information shared will be different today because students are at another stage of the problem. If we have a time line in place or a goal of finishing the Project Problem by a certain date, I give students reminders.

2. Work Session

Again, I circulate through the room, observing, discussing, note-taking, and assisting.

V. Day 5 - Closure

This is usually the last day, but I try to be flexible. If more work days are needed, we take them.

A good way to bring closure to a Project Problem is to have groups do presentations of their products and/or solutions. This gives students an opportunity to discuss their products, solutions, techniques, and ideas for solving. If you require a written response such as a math journal entry, today is a good day to have students begin writing while so much summary thinking and discussion is in the air. Outside payoffs (like the aliens from Zignox making good on their promise of "richly rewarding" students) usually take place today.

Students' work on a Project Problem often results in products that beg to be displayed. We have used hallway wall space, our media center, and display cases to exhibit our final products, written work, photographs, and written reactions. Students are very proud of their intensive and prolonged efforts. It is fun for them to let the world at large (or at least the school) see what they have accomplished. Inviting parents, administrators, and media in to see the completed work can also be valuable. It was a big undertaking, and it is now a time to celebrate.

Chapter 2 Notes
[1] *Structuring Cooperative Learning: The 1987 Lesson Plan Handbook* by D.W. Johnson, R.T. Johnson, and Edythe Johnson Holubec. Edina, MN: Interaction Book Company, 1987, p. 17. An excellent source for making decisions about setting up cooperative learning groups.
[2] *Structuring Cooperative Learning: The 1987 Lesson Plan Handbook*, p. 16.

Chapter 3 - The Project Problems

INTRODUCTION TO THE PROBLEMS

There are two problems for each grade level, Kindergarten through Grade 3. They concentrate on major math concepts common to curricula at these grade levels. Depending on the ability of your particular class, however, you may want to try problems designated here for a different level. These grade level designations are not etched in stone; rather, they are approximations or suggestions. In addition, you will find that altering certain parts of a problem can easily change a first-grade problem into a second or a third into a fourth. Each problem has a "Customizing the Project" section which should help in this area. Feel free to adapt any problem and make it your own.

TRYING OUT PROBLEMS

If I could recommend just one thing to do to help ensure success with these problems, it is this: Try these problems yourself first. Do them as the students are required to do them. You will then know firsthand the processes needed, the pitfalls, and the successful strategies your class will most likely encounter. You will be able to anticipate and prepare for your students' needs before they even get near the problem.

The most enjoyable way to do this is with other teachers. Enlist a small group of teachers, preferably at your grade level. Have each person work through the problem individually or work it together as a small group. The insight you gain will help you to brainstorm how best to implement and assess the problems. Jump in and have fun!

Dear Students,

My name is Shapey, and I live on a planet far away in outer space. I have heard that you earthlings are good problem solvers and that you are also very nice and helpful. Well, I have a problem and need your help.

My mother's birthday is coming soon. I would like to give her a portrait of me as a birthday present. I cannot make the portrait because I am not very artistic. Could you please help me by making the portrait?

Here is some information to help you make my portrait.

- I am 10 inches tall.

- I am 8 inches wide.

- People say I'm very "shapely."

  I am made of   4 rectangles,

  2 circles,

  5 triangles,

  and 1 square.

- People say that my 2 stop-sign-shaped eyes are cute.

- I have the same 5-sided nose as my father.

- I have my mother's crescent smile.

- I am symmetrical.

Could you also put my portrait in a frame? I think that my mom would really love a frame decorated with a pattern that uses at least 3 different shapes.

Good luck and thank you for all your help. I can't wait to see my portraits. I'll be visiting soon and will say thanks in my own special way.

Your friend from Space,

SHAPEY

SHAPEY

Grade Level: Kindergarten

Overview

A letter arrives from outer space. It is from a strange, but friendly, alien named Shapey. His mother's birthday is coming up soon, and he wants to give her a picture portrait of himself as a gift. He needs help to make it. Students use their artistic and mathematical skills creating and manipulating shapes to make portraits according to the description Shapey provides.

This problem can be made appropriate for students in Grades 1 and 2 as well. See Customizing the Project, page 17.

Objective

- To explore measurement, symmetry, and basic geometric shapes in a problem solving context

The Mathematics

- *identifying geometric shapes*
- *creating geometric shapes*
- *counting*
- *standard measurement*
- *symmetry*
- *patterning*

Time Needed   four to five 40-minute periods

Grouping   This problem is very good for individual independent work, but it can be done by partners or in small groups. There are enough tasks to keep a small group active if the workload is divided among group members.

Preparation and Materials

Day 1

- Shapey letter, 1 per student, page 19 (Make individual letters and/or copy the letter onto a class chart for group reading.)
- Large envelope decorated with outer-space art in which to deliver the Shapey letters (You may instead wish to create a "flying saucer" to hold the letters. Do this with two pie plates. Place folded letters in one plate and place the other plate upside down on top of the first to create a saucer shape. Cover the saucer with foil and decorate. If you choose to have the letter on a class chart, the chart can be rolled up and delivered to the class in a decorated mailing tube or simply rolled up and secured with a rubber band.)
- Sample of a decorated frame (optional)

Days 2, 3, and 4

- Colored construction paper, 1 ream of assorted colors
- Crayons, 1 set per student
- Scissors, 1 per student
- Glue, as needed
- Large white construction paper, 1 sheet per student

- Rulers, 1 per student
- Templates, such as Attribute Blocks, for tracing shapes (optional)

Day 5

- Thank-you notes, 1 per student, page 20
- Thank-you shapes, 1 per student (You can make these from any material that is handy, but colored metallic paper seems to enchant young students and has an "outer-spacey" feel to it. Place the shapes in the thank-you note.)

PROJECT: SHAPEY

Day 1: Special Delivery

Deliver

"Discover" a mysterious, strangely decorated envelope (or mailing tube or flying saucer) in the schoolyard and bring it into the classroom. It might be fun to have an accomplice (perhaps another teacher or a former student who is now in an upper grade) discover the envelope outside and bring it up to your window. Or, you may want to return from lunch or gym and find it lying in the middle of the classroom floor. Open it slowly and carefully. Remember that it may be from outer space. Ask students what they think it might be. Distribute the letters to students or display the chart letter.

Read

Read the letters or the chart letter aloud to the class. Students who are readers can join in and help read aloud as much as possible. Discuss vocabulary and terminology that might be new to students.

Discuss

The problem is woven into the context of the letter. To help students interpret the letter and understand the task, ask

1. So, what's the problem?

Use some or all of the following questions and narrative to guide the discussion and analysis of the letter.

> *Who wrote you the letter?* (Shapey, a friend from outer space)

> *Why did he write to us?* (He wants our help in making portraits of him for his mother.)

You may want to show some examples of portraits.

> *What is a portrait?* (It is picture of a person. It could be the person's face, whole head, body from shoulders up, or whole body.)

> *So, he wants us to make a picture of him. What did he say he wanted to put the picture in?* (A frame)

Hold up a frame or describe for students.

> *It might be fun to make a picture of Shapey. How big should it be? Did he say how big he is?* (Yes, he said he is 10 inches tall and 8 inches wide.)

> *How big is that? Can you show with your hands? How do we know?*

This could be a good opportunity to do a mini-lesson on measurement with inches and rulers. The problem has supplied students with a real reason and an immediate need for this information. You can present the mini-lesson beforehand to prepare students for dealing with the problem, or you can present it as they work through the problem.

> *What else did Shapey tell us about himself? Why is he called Shapey?* (He is made of lots of shapes. He is made of 4 rectangles, 2 circles, 5 triangles, and 1 square. He also has 2 eyes shaped like stop signs, a 5-sided nose, and a crescent smile.)

> *What do you know about shapes?* (Allow time for students to share and discuss what they know to help build a connection to previous knowledge and make the problem more motivating.)

You may want to discuss the shape names, *octagon* and *pentagon*, if appropriate for your class.

This is an opportunity for a discussion of shape concepts if introduced previously or an opportunity for a mini-lesson, either now or as the problem solving process continues. The motivator, again, is that this is information students need and want.

> *I noticed Shapey said he is symmetrical. Does anyone know what that means?* (It means both sides of Shapey are the same and even, pretty much the way we are.)

This is a another mini-lesson opportunity. Paper folding or looking at partners can help bring across the idea of symmetry.

> *Thinking about the frame again, does Shapey want some kind of special frame?* (Yes, it should be decorated with a pattern that uses at least 3 different shapes.)

> *What does Shapey say at the end of the letter?* (He would like to visit us to see the portraits and thank us in his own special way.)

> *There is a lot to remember here. I think I will post the information (leave the chart up) for everyone to see while we work, so it can remind us of what Shapey has asked us to do.*

Once you have finished reviewing the information in the letter, ask

2. What materials do you think you will need to solve the problem?

List responses on the board. Ask students to justify their thinking regarding the need for each material. This process helps students to think about strategies and approaches. Assure students that all necessary materials will be ready for them the next day or whenever the Day 2 session will take place.

Day 2: Let's Get to Work!

Have all materials ready for students. Briefly review the questions and information from Day 1 so everyone is clear on the task and parameters. Have students begin work. Circulate, assist, and observe.

Common Student Approaches

Students often begin by making the shapes and pasting them down on background paper. It is a good idea to remind those who choose this approach that they will need to think about measurement at some point early on in their work. How to make the shapes fit in the measured space is an important part of their discovery learning. Students may want to use stencils or shapes to trace from as they create their shapes.

Rather than tracing and cutting out shapes, some students will choose to draw the shapes. In this case, it is common for these students, particularly older ones, to produce rough drafts and then refine the work in a final draft.

Teacher Role

While students are working, circulate, assist, and observe. In addition, use this time to begin assessment. Keep these assessment questions in mind:

What math are students using?

What organizational ideas are students using?

What tools are students using?

What are students doing well?

What difficulties are students having?

Day 3: Time to Share

Begin with a whole class discussion to give students the opportunity to think about their partial solutions in light of what their peers share. Have students sit in a circle with their work. Ask the following questions and discuss answers together:

What did you (or your group) do first? (Let students discuss pros and cons of various suggestions.)

What has worked well for you?

What problems have you had?

Has anyone used a ruler yet? Why or why not?

How do you use a ruler? Who can demonstrate?

How did you make your shapes?

How do you know how many shapes you have made?

How do you know which shape is which? Describe the shapes you've made.

How do you keep track of what you are doing? How do you know what has been done?

Did anyone start on the frame yet?

What are some of the patterns we have so far on frames? (Let students share pattern ideas.)

After the discussion, direct students to continue working. Again, as you did yesterday, circulate, assist, observe, and assess.

Day 4: Still Working

Begin with a whole class discussion and share session. Have students bring up all work in progress. Allow students to share their work so far. Ask the following questions and discuss:

What has worked well for you?

What problems have you had? How are you trying to solve these problems?

After discussion and shared answers, have students return to work. Continue to circulate, assist, observe, and assess.

Have all students display their portraits. Leave them on display so Shapey can come and visit some time when students are out of the room. When students return to class, they will find evidence that Shapey has visited in his thank-you note and the "thank-you shape" he has left each student.

You may want to take photos of the portraits.

Use these questions to help summarize the week's work:

> *Was there any one right way to make Shapey? Explain.*

> *Do we know exactly what Shapey really looks like? Why or why not?*

> *He did need to have certain things in his portrait though, didn't he? Like what?*

CUSTOMIZING THE PROJECT

This can be a very challenging problem for many kindergartners. Consider eliminating certain parameters to best suit the needs and abilities of your group. Do try, however, to retain some aspects of both measurement and basic shapes so students can have the experience of manipulating both in the same problem environment.

The Shapey letter has a number of parameters for creating the portrait. These can be easily altered to meet the needs of various age and ability levels. For example, the measurements can be shifted from standard to non-standard units; the type or number of shapes used can be changed; colors can be added to the description; the concept of symmetry can be eliminated.

The other kindergarten problem in this book, *A Very Special Zoo*, also explores patterns, and measurement. Shapey explores some of the same elements but in a more directed fashion. Depending on your group, your style, and what you are working towards with them, *Shapey* may be a good follow-up, precursor to, or alternative to *A Very Special Zoo*.

ASSESSMENT

Review completed portraits. Consider these questions as you review the final product and the process you observed over the course of the project:

For additional tips on assessment see Appendix 1, Assessment, pages 97-99.

> *How do Shapey's height and width compare to the requirements?*

> *How do the number and type of shapes used compare to the requirements?*

> *Is Shapey symmetrical?*

> *Does the frame have a consistent pattern using at least three different shapes?*

> *How complex is the pattern? Is the complexity of the pattern something you want students to work towards?*

> *To what degree did the student participate in the process? How much assistance was needed from the teacher?*

Math Journal Writing or Interviews

Ask students to answer one or more of the following questions either in writing or orally:

> *What math ideas did you use to make Shapey?*

How did you find out how big to make Shapey?

What is symmetry?

Draw a triangle (rectangle, circle, square). How do you know it is a triangle (rectangle, circle, square)?

What is one good way to start to make Shapey? What would you do first if you did it again? Why?

CURRICULUM CONNECTIONS

Literature *Space Case*, by James Marshall, is a good picture book to read aloud with this problem. A strange visitor from outer space is featured in a colorful story. The artwork is very "shapely."

Math Have students explore further the idea of symmetry. One way to do this is to have one student use Attribute Blocks to create "half of a design" along the edge of a ruler while a partner tries to complete the design in a symmetrical manner.

Art Share portraits done by noted artists. Show students a variety of styles and types. Discuss self-portraits. Supply mirrors and let students try to make self-portraits.

Music Listen with students to Hap Palmer's song about *Shapes*.

Language Arts Have students write or dictate a story about Shapey. Ask students to imagine what his planet is like and what he likes to do there.

Social Studies Discuss these ideas with students: Shapey asked for our help and we helped him by making portraits. Are there other ways we can help people we know? Are there people who help us? Who are they? How do they help us?

Science Ask students to think about our planet. Ask them to name the planet on which we live. Have students describe some aspects of Earth that Shapey might be interested in knowing about.

Dear Students,

My name is Shapey, and I live on a planet far away in outer space. I have heard that you earthlings are good problem solvers and that you are also very nice and helpful. Well, I have a problem and need your help.

My mother's birthday is coming soon. I would like to give her a portrait of me as a birthday present. I cannot make the portrait because I am not very artistic. Could you please help me by making the portrait?

Here is some information to help you make my portrait.

- I am 10 inches tall.
- I am 8 inches wide.
- People say I'm very "shapely."

  I am made of 4 rectangles,

  2 circles,

  5 triangles,

  and 1 square.

- People say that my 2 stop-sign-shaped eyes are cute. ⬡ ⬡
- I have the same 5-sided nose as my father. ⬠
- I have my mother's crescent smile. ⌣
- I am symmetrical. ⚥

Could you also put my portrait in a frame? I think that my mom would really love a frame decorated with a pattern that uses at least 3 different shapes.

Good luck and thank you for all your help. I can't wait to see my portraits. I'll be visiting soon and will say thanks in my own special way.

Your friend from Space,

SHAPEY

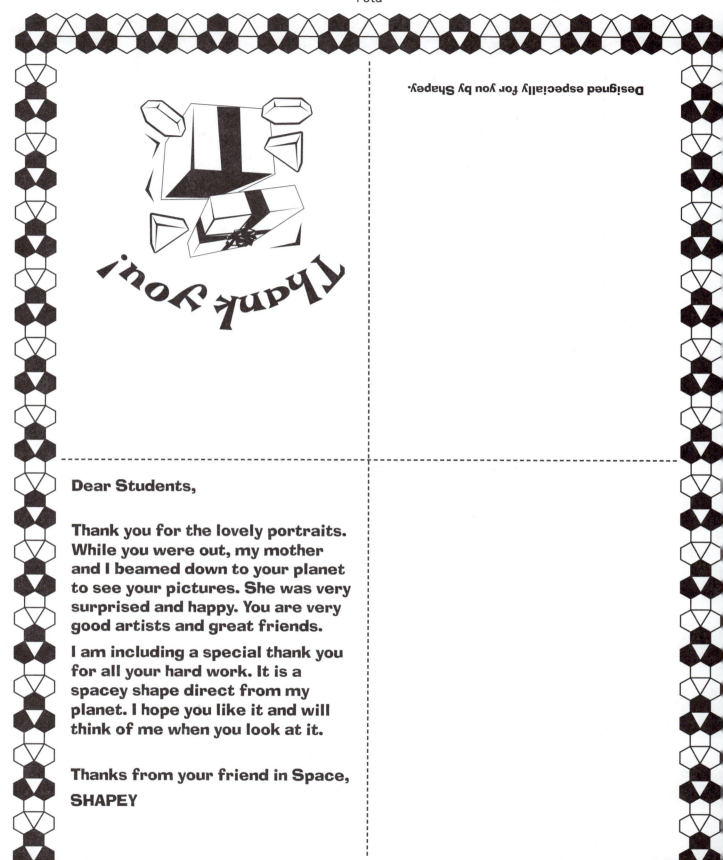

Designed especially for you by Shapey.

Thank you!

Dear Students,

Thank you for the lovely portraits. While you were out, my mother and I beamed down to your planet to see your pictures. She was very surprised and happy. You are very good artists and great friends.

I am including a special thank you for all your hard work. It is a spacey shape direct from my planet. I hope you like it and will think of me when you look at it.

Thanks from your friend in Space,
SHAPEY

Place a shape in lower right quadrant. See Preparation and Materials, Day 5, for instructions.

A Very Special Zoo

Dear Boys and Girls,

I am so sad. Maybe you can help me. Let me tell you my story. I lived in a zoo on a tiny island in the Atlantic Ocean. All my animal friends were special like me. We were all part one animal and part another. I am part zebra and part giraffe, and so I am called Zee-Raffe.

I loved that zoo. My animal friends and I each had our own specially decorated cages. The cages were connected by beautiful patterned paths.

One night I sneaked out of the zoo and went for a midnight swim. I swam too far. The next thing I knew, I was on a beach in the U.S.A. During the day I slept in different hiding places. At night, I traveled. I have been to mountains, deserts, and forests. Now I am here.

I really like your school. I would like to stay here, but I do miss my friends and I do miss my old zoo.

Can you help me feel more at home here? One way you could do this is to make me pictures of new animal friends like me that are part one animal and part another. Could you also make cages for the new animals and connect them with patterned paths? I would be so grateful if you could do this for me.

Thank you,

Your new friend,

ZEE-RAFFE

A VERY SPECIAL ZOO

Grade Level: Kindergarten

Overview

Students receive a visit from a very unusual animal. He is part zebra, part giraffe – a zee-raffe! The letter around his neck says that he is lost and sad. He misses his friends (who are all part one animal and part another) and the special zoo he came from. He asks students to work together to create pictures of other "part and part" friends and a new zoo to help cheer him up and make him feel at home.

Objective

- To explore and apply counting and nonstandard measurement in a problem solving context

The Mathematics

- *counting*
- *estimation*
- *patterning*
- *1-to-1 correspondence*
- *constructing sets of 10*
- *ordering and matching numbers through 10 and beyond*
- *nonstandard measurement*
- *number recognition*

Time Needed   five or more periods, each 40 to 60 minutes

Grouping   Partner groups are ideal.

Preparation and Materials

Day 1

- Zee-Raffe (You can make this part-zebra, part-giraffe creature from paper mache, clay, wood, a combination of stuffed animals, or any other material. Once the zee-raffe is made, it can be used again and again. If creating a model is not possible, you can draw a picture of the "part and part" animal or make a two-dimensional picture by cutting animal pictures from magazines and gluing them together.)
- Letter from Zee-Raffe, page 29 (Copy the letter onto chart paper and string it around the creature's neck.)
- Chart paper, 2 sheets
- Marker for chart

Days 2 and 3

- Parent volunteers (Optional; some kindergarten students may need assistance with the parts of this problem that involve measuring and counting with cubes, cutting out, and taping. Having parents available to help out will enable you to circulate more freely and do more direct supervision.)
- Drawing paper, 1 sheet per student

> This can be an engaging problem for first graders as well. See Customizing the Project, page 27.

> Making Zee-Raffe can be an interesting project for older students in art classes or for a school art club that is looking for a fun service project.

- Glue, 1 bottle for every two groups
- Tape, 1 roll for every two groups
- Markers, 1 set per group
- Crayons, 1 set per group
- Animal reference books, as needed

Day 4

- Parent volunteers (optional)
- Large sheets of colored craft paper, 1 per group
- Snap™ Cubes (or Unifix® Cubes), 2 class sets
- Tens strips based on Snap Cubes, about 15 strips per group, page 30
- Scissors, 1 per group
- Markers, 1 set per group
- Crayons, 1 set per group

Day 5

- Strips of paper to create a path, several per group
- Materials (such as rubber stamps and precut shapes) for making path patterns
- Stamp pad, 1 for every two groups
- Glue, 1 bottle for every two groups
- Markers, 1 set per group
- Crayons, 1 set per group
- Thank-you card from Zee-Raffe, page 31

PROJECT: A VERY SPECIAL ZOO

Day 1: Special Delivery

Deliver

Arrange for the special animal to arrive at the classroom in some dramatic manner. One way to do this is to have an accomplice place the animal outside the door at an opportune time. Bring the animal model or picture into the room. Ask students what they think the creature could be. Let students discover the chart letter around its neck. Display the chart letter.

Read

Read aloud the chart letter to the class. Discuss vocabulary and terminology that might be new to students.

Discuss

The problem is woven into the context of the letter. To help students interpret the letter and understand the task, ask

1. So, what's the problem?

Use some or all of the following questions to guide the discussion and analysis of the letter.

Why is Zee-Raffe so unhappy? (He misses his friends and his zoo.)

How many zoo animals can you name? (Brainstorm a list and record responses on a chart. Allow time for students to share what they know about animals. Help students differentiate between zoo animals and house pets.)

What do Zee-Raffe's animal friends look like? (They are all part one animal and part another animal.)

What was Zee-Raffe's zoo like? (Each animal had a cage, and the cages were connected by patterned paths.)

What can we do to make Zee-Raffe happy again? (We can create pictures of his animal friends and a picture of a new zoo.)

Once you have finished reviewing the information in the letter, ask

2. What materials do you think you will need to solve the problem?

List responses on the board. Ask students to justify their thinking regarding the need for each material. Tell students you will gather necessary materials and have them ready for work the next day.

This discussion about animals helps build a connection to previous knowledge and makes the problem more motivating.

Day 2: Let's Get to Work!

Have all materials ready. Form partner groups. Briefly review the questions and information from Day 1 so everyone is clear on the task and parameters.

Introduce the set of animal books that will serve as resources for students in both their thinking and their drawing. Remind students of the chart list of animal names that was brainstormed on Day 1. Assure students that they can practice drawing and do rough drafts as needed. Then have students begin work.

Common Student Approaches

When student s work in pairs, often one strong leader will emerge. It is important to emphasize to students that the outcome must be pleasing to *both* partners. This can become an issue early on when students begin discussing who will do the front or back of the animal.

As students work on creating cages, the counting of larger numbers of cubes can be difficult. Often though, the concept of grouping cubes by tens will arise as students try to organize the larger amounts of cubes. You may want to share this with other groups as it occurs or help students move toward this idea as a good strategy.

Teacher Role

While students are working, circulate, assist, and observe. In addition, use this time to begin assessment. Consider these assessment questions as you observe and interact:

What math are students using?

What organizational ideas are students using?

What tools are students using?

What are students doing well?

What difficulties are students having?

Some teachers like to organize this activity as a center. Students are placed into working groups so that two or three partners may be working on the project while the rest of the class is working at other centers. Groups then rotate through the problem over the course of a week or two.

Day 3: Time to Share

Begin with a whole class discussion to give students the opportunity to think about their partial solutions in light of what their peers share. Have partners sit in a circle with their animal pictures displayed in front of them. Ask the following questions and discuss answers together:

What did you do first? (Discuss pros and cons of various suggestions.)

What has worked well for you?

What problems have you had?

After the discussion and shared answers, have students return to work. Again, as you did on Day 2, circulate, assist, observe, and assess.

Day 4: Still Working

Most animals will be completed during Days 2 and 3. Begin with a whole class share session and shared display of models. Ask the following questions and discuss the answers:

Now that you have completed the animals, what did Zee-Raffe ask you to make for them? (Cages)

How will you find out what size cage your animal needs? (Measure)

Tell students that you would like them to use Snap Cubes (or Unifix Cubes) for the measuring because you have special "cage-building" materials that work well with the cubes. These materials are the tens strips.

Give each partner group a large piece of precut craft paper. Ask them to use the cubes to create the top, bottom, left side, and right side of the cage for their animal. Ask groups to count the number of cubes they used for their cage.

As groups complete their "cube" cages, distribute the tens strips. Ask students to recreate the size and shapes of their cages with these. For example, if one side of the cage measures 42 cubes, students would lay out and paste down four strips of ten squares and then cut out two additional squares and paste those down. Then the students would write 42 on that side of the cage. When all four sides of the cage are pasted in place and labeled with a number, students should cut away the excess craft paper that is outside the cage.

As students work on their cages, continue to circulate, assist, observe, and assess.

Day 5: Closure

Have each group display their finished animal and cage. Allow time for students to explain about their animal and to discuss the number of cubes that form the cage and the ways that the cubes can be counted. Have each partner group decide on a name for their creature and make a label for its cage.

Reread the letter aloud again and ask

Is there anything else we need to add to the zoo? (Yes; beautiful patterned paths between the cages)

Patterned paths can be created in a variety of ways. Students can use the tens strips or any other type of path as the basis for creating a pattern. Have a variety of materials available for creating patterns, such as markers, rubber stamps, and precut shapes. Display the unusual creatures in their cages connected by beautiful patterned paths on the walls of the classroom or the hallway.

Inviting other classes for a tour of the zoo, complete with zoo refreshments like popcorn and lemonade, is a closure activity that is lots of fun. You may also want to send your students the thank-you note from Zee-Raffe.

CUSTOMIZING THE PROJECT

There are three components students must create in this problem: the animal, the cage, and the patterned paths. You can eliminate the path requirement to make the problem quicker and simpler. You can lower the number of cubes used by providing students with smaller pieces of drawing paper, thereby restricting students to making smaller animals that require smaller cages.

The problem can be made more complex by having students create three-dimensional animals. This would necessitate making three-dimensional cages which are more challenging to build and measure. Some first graders or even older students may be interested and ready for a three-dimensional version of the problem.

ASSESSMENT

Examine each model for the required criteria and think about the process and your observations. Consider the following questions:

Is the animal part one animal and part another?

Is there a cage that fits around the animal?

Was the student able to count 10 cubes? More than 10 cubes?

Was the student able to construct a set of 10 cubes?

Was the student able to make a strip that corresponds to the length of a set of cubes?

Was a pattern created and extended in the path?

For additional tips on assessment, see Appendix 1, Assessment, pages 97-99.

Math Journal Writing or Interviews

Ask students to answer one or more of the following questions either in writing or orally:

How did you make decisions? How did you cooperate with your partner?

How did you measure the cage?

What is a good way to count more than 10 cubes?

What is a pattern? Can you give an example?

CURRICULUM CONNECTIONS

Literature Reading aloud *Wing Ding Dilly* by Bill Peet and *Zoo Looking* by Mem Fox to students will introduce them to some unusual and interesting animals.

Math Have students use Multilink™ Cubes to create models of animals and count the number of cubes used.

Art Help students to add to the wall display by having them draw features found in zoos, such as trees, people, balloons, vendors, benches, and signs.

Music Listen with students to zoo songs such as Peter, Paul, and Mary's "Going to the Zoo" from the album *Peter, Paul, and Mommy* and "Going to the Zoo" from Raffi's *Singable Songs for the Very Young.*

Language Arts Work with students to write an invitation to their families to come for a tour of the zoo. Have students practice being tour guides and prepare explanations of their zoo to visitors.

Social Studies Map skills can be introduced as the patterned paths are followed around the zoo from creature to creature. Work with students to create a map of the zoo.

Science Read aloud and have students look at books about animals. Have students look carefully at the similarities and differences between animals and discuss these attributes in light of the animals they created in the problem.

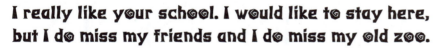

Dear Boys and Girls,

I am so sad. Maybe you can help me. Let me tell you my story. I lived in a zoo on a tiny island in the Atlantic Ocean. All my animal friends were special like me. We were all part one animal and part another. I am part zebra and part giraffe, and so I am called Zee-Raffe.

I loved that zoo. My animal friends and I each had our own specially decorated cages. The cages were connected by beautiful patterned paths.

One night I sneaked out of the zoo and went for a midnight swim. I swam too far. The next thing I knew, I was on a beach in the U.S.A. During the day I slept in different hiding places. At night, I traveled. I have been to mountains, deserts, and forests. Now I am here.

I really like your school. I would like to stay here, but I do miss my friends and I do miss my old zoo.

Can you help me feel more at home here? One way you could do this is to make me pictures of new animal friends like me that are part one animal and part another. Could you also make cages for the new animals and connect them with patterned paths? I would be so grateful if you could do this for me.

Thank you,

Your new friend,

ZEE-RAFFE

Tens Strips

Fold

Dear Girls and Boys,

Thank you so much for all your help. I love my new zoo and new friends. I am so happy now, thanks to you.

Your friend,

ZEE-RAFFE

Dear Students,

I am Roberta Recycle, Queen of Recycling Robots. We are special robots made out of recycled materials. Our job is to remind everyone to recycle. I come to you for help. We all know that pollution and too much trash are problems. But we can do something about it. We can let people know about recycling.

Please help me spread the word about recycling by creating a special helper – your own Recycling Robot. A Recycling Robot is a robot made out of juice boxes, tape, and recycled and scrap materials like bottle caps, straws, and yarn.

To make a Recycling Robot, there are certain rules you must follow:

1. It must be built so that it is very sturdy and can be moved easily from place to place.
2. It has to hold a sign with a message to remind people to recycle.
3. It must be built of Robo-Blocks. A Robo-Block is 10 juice boxes put together with tape to make a strong, sturdy block of any shape.
4. A Recycling Robot must be made of at least 10 Robo-Blocks. You may use more.
5. A Recycling Robot must have a name.

We would like each of you to make a Robo- Block from 10 juice boxes. Then, on the Robot Recording Sheet, draw a picture of the Robo-Block you made. Write a number sentence that matches your Robo-Block and answer the three questions. Give this to your teacher as a record of your work. Finally, combine all the Robo-Blocks in your group to create a Recycling Robot.

Good luck and happy recycling!
Thanks for your help,
ROBERTA RECYCLE,
Queen of the Recycling Robots

RECYCLING ROBOT

Grade Level: 1

Overview

Students get a letter from Roberta Recycle, Queen of Recycling Robots, requesting their help in creating a Recycling Robot to help inform people about recycling. Each student uses 10 juice boxes or milk cartons to create a Robo-Block and to explore representations of 10 as physical models, pictures, and number sentences. Students in each group work together to combine their Robo-Blocks to create a Recycling Robot.

Objective

- To investigate and make connections between ways to make 10 using blocks, pictures, and number sentences

The Mathematics

- *making 10s*
- *counting (by 1s, 5s, 10s)*
- *addition*
- *grouping by 10s*
- *multiplication*
- *concept of 100*

Time Needed   four to five 40-minute periods

Grouping   This problem works best with 10 students in a group, but a larger group is possible.

Preparation and Materials

Day 1

- Roberta Recycle's letter, 1 per student, page 39 (In addition to having copies on letter-sized paper, you may want to copy the letter onto chart paper to make it easier for all to read together.)
- Robot Recording Sheets attached to letters, 1 per student, page 40
- A used (recycled) envelope or one made of used paper in which to deliver the letters and recording sheets (If you choose to use a chart letter, you can roll up the chart with the letters and recording sheets and put them in a used mailing tube or simply roll them up and secure with a rubber band.)
- Pre-made sample of a Robo-Block

Days 2, 3, and 4

- Juice boxes or milk cartons, 10 per student (Juice boxes need to be cleaned thoroughly. Clip off one top corner, flush with water, and let dry. Milk cartons are more problematic in that the peaked tops need to be flattened. Any small boxes or containers of uniform size can be substituted.)
- Masking tape, 1 to 2 rolls per group
- Scissors, 1 per student
- Decorative recycled junk for making the robots truly distinctive (Optional; a

This problem can be made appropriate for students in Kindergarten or in Grade 2. See Customizing the Project, page 37.

A good way to collect boxes is to make announcements at the school cafeteria and put up garbage bags to collect them. Requesting juice boxes from home is another possibility.

few weeks before you begin this project, start collecting materials such as bottle caps, straws, and bits of foil and yarn.)

- Glue, 1 bottle per group
- Poster board for presentation chart (optional)
- Markers for presentation chart (optional)

Day 5

- Thank-you notes from Roberta Recycle, 1 per student, page 41

PROJECT: RECYCLING ROBOT

Day 1: Special Delivery

Deliver

Arrange for delivery of the letters and recording sheets to students. If you have time or help, an interesting way to deliver these is to fold up each student letter and recording sheet, insert them into an empty juice box, and leave a box on each student desk. Even if you use a chart letter, be sure each student gets a copy of the letter and attached recording sheet. If you are using an envelope or mailing tube, open it and pass out letters and/or display the chart.

Read

Give students time to read the letter or chart silently. Then have them take turns reading aloud or read aloud to them as they follow along. Discuss any vocabulary and terminology that might be new to students.

Discuss

All of the information needed to understand the problem is in the letter. During discussion, it is important that students get a good understanding of just what the problem is and what they are required to do. Ask

1. So, what's the problem?

Use some or all of the following questions to focus the discussion and help students understand the letter and the problem.

Who wrote you the letter? (Roberta Recycle, Queen of Recycling Robots)

What does recycling mean? (Allow time for students to share what they know about recycling. This helps build a connection to previous knowledge and makes the problem more motivating.)

What does Roberta want your group to do? (Build a Recycling Robot to help spread the word about recycling)

What is a Recycling Robot? (A robot made out of juice boxes, tape, and recycled and scrap materials)

At this point, it is a good idea to display a sample Robo-Block.

The letter says the robot must be built out of Robo-Blocks. What is a Robo-Block? (A Robo-Block is 10 juice boxes put together with tape to make a strong, sturdy unit block of any shape.)

How many Robo-Blocks are used to build a Recycling Robot? (10 or more)

Does the letter say anything else about how the robot should be built? (It has to be sturdy yet easy to move from place to place.)

Does the robot have to hold anything? (Yes; a sign with a message about recycling)

Why are we building a Recycling Robot? (To help advertise the message about recycling)

Do you think it will be important to cooperate and plan as a group? Why or why not? (Yes; a lot of people are working together to build one object.)

What else do you have to do besides build the robot? (Fill in a Robot Recording Sheet and give it to the teacher; name the robot.)

When you think that students have all the information they need and a basic understanding of the problem and letter, ask

2. What materials do you think you will need to solve the problem?

List responses on the board. Ask students to justify their thinking regarding the need for each material. This process helps students to begin thinking about strategies and approaches to solving the problem. Assure students that these materials will be made available to them when they begin.

You may wish to guide students through the sample Robo-Block picture and number sentences on the recording sheet.

Day 2: Let's Get to Work!

Have all materials ready for students. Assign students to groups. Briefly review the questions from Day 1 so everyone is clear on the task and parameters. Remind students to complete a Robot Recording Sheet as they work. Before students begin work, ask

Are there any questions or concerns about the problem before we begin?

Do you all know what you need to do?

Think about what part of the robot your Robo-Block will be. Will it be an arm? the head? You may want to talk to your group members and plan out some ideas before you begin to build.

Once you are satisfied that students have a reasonable understanding of the task, have them begin work.

Common Student Approaches

Some students will spend a great deal of time on the accuracy of their drawings; others will rush to build. Remind the "rushians" to be sure that the pictures, number sentences, and blocks all correspond with each other on their Robot Recording Sheets and that they are readable.

A big aspect of this problem is cooperation among students in dividing up responsibilities, particularly when it comes to the assembly of the robot. Observe groups early on to see if students have devised ways to share the work load.

Teacher Role

While students are working, circulate, assist, and observe. Use this time to begin assessment. Consider these assessment questions:

What math are students using?

How is the group cooperating and dividing up tasks?

What strategies are students using?

How efficient are the strategies?

What tools are students using?

Are students using 10 boxes to build their Robo-Blocks?

Are students recording their Robo-Block information on the recording sheets?

How is counting being done? By 1s? 5s? 10s?

What are students doing well?

What difficulties are students having?

Day 3: Time to Share

Begin with a whole class share session. Have groups sit in a circle with their work displayed in front of them. Encourage students to listen for and share ideas during discussion. Ask the following questions and discuss answers together:

What did you do first?

How did you cooperate?

How did you decide who would do what part of the robot?

Have you thought about how you will work together to join your Robo-Blocks into one big robot?

What has worked well for you?

What problems have you had?

During discussion, take advantage of the opportunity to reinforce valuable concepts or ideas students suggest. After discussion and shared answers, have students begin work. Again, circulate, assist, observe, and assess.

Only so many students in a large group can assemble the Robo-Blocks into the robot. Encourage students to take turns working in smaller groups of two's or three's to assemble their parts. Those who are not assembling can complete and check their Robot Recording Sheets. Another option is to ask the nonassemblers in each group to collaborate on a presentation chart showing a picture of their robot with an arrow pointing to the part created by each student and the corresponding number sentences for each part.

Day 4: Still Working

Begin with a whole class share session and a shared display of robots and Robot Recording Sheets. Ask the following questions and discuss:

What has worked well for you?

What problems have you had?

What have you done about these problems?

After discussion and shared answers, have students begin work again. As groups complete actual construction of their robots, they may want to begin decorating at this time. For example, they may attach bottle caps for eyes, yarn for hair, and so on.

As before, circulate, observe, assess, and assist as necessary.

Day 5: Closure

If groups have prepared a presentation chart, have them bring this and their robot to the front of the class. If there is no chart, have the groups bring up their recording sheets and robots. Have each group tell briefly about how they organized themselves and cooperated. Use questions like these to guide the discussion:

How did you decide who would do what part of the body?

How are the Recycling Robots the same? How are they different?

Why do the Recycling Robots all look different? Was there any one right way to make the robot?

After presentations and discussion, Recycling Robots can be displayed around the school or be sent on visits to different classrooms or community sites (local public library, shopping mall, municipal building) to spread the message about recycling.

Arrange for the thank-you letters from Roberta, Queen of Recycling Robots, to be delivered. Read them aloud together. Sometimes Roberta likes to reward each student with a new, full, juice box as a thank-you for a job well done.

CUSTOMIZING THE PROJECT

The total number of juice boxes, Robo-Blocks, and robot body parts can all be adjusted to make the problem easier or more complex. The whole problem can be altered so that each group creates one or two robot body parts instead of a whole robot. In that case, each group would need to create specific robot body parts and then work with other groups to assemble one class robot.

A scaled-down version for kindergartners might involve giving each student 10 juice boxes to create one individual robot. Students can then bring their robots together to count by tens and find the total number of juice boxes used.

Second graders can begin to look at, describe, and write about Robo-Blocks as fractional parts of the completed robots.

ASSESSMENT

Review individual Robot Recording Sheets and the finished robots. Consider these questions as you look at the final product and reflect on the process you observed over the course of the project:

Does the picture match the number sentence and actual Robo-Block?

What kind of number sentences were generated? Simple addition facts? Repeated addition? Multiplication?

Is the number of blocks on the Robot Recording Sheet accurate?

How was the counting of the total number of boxes done? By 1s? 2s? 5s? 10s?

Does the robot have an appropriate sign?

Is the robot sturdy?

Is it easily moved?

For additional tips on assessment, see Appendix 1, Assessment, pages 97-99.

Math Journal Writing or Interviews

Ask students to answer one or more of the following questions either in writing or orally:

What math did you use to solve the problem and complete the robot?

Is there more than one number sentence for a Robo-Block? Write down the sentences used by your group for your robot.

How many juice boxes were in your completed robot? How did you count them?

If you were to share one strategy or technique that worked particularly well for you in solving this problem, what would it be?

CURRICULUM CONNECTIONS

Literature This problem can be an outgrowth of a recycling or pollution study or theme. Talk to students about what they can personally do about recycling and pollution. A good book to refer to is *Ecology for Every Kid* by Janice Van Cleave.

Math Make a chart showing the number sentences the class came up with for 10. Challenge students to create even more.

Art Have students do a painting or drawing of their group robot or take time to decorate their robot further with recycled accessories.

Music Some music sounds very mechanical or robot-like. See if students can find any music that reminds them of robots.

Language Arts Invite students to write a speech for their robot to give about recycling. Tape record the speech. Put a tape player behind the robot. Turn it on and let the robot give the speech.

Social Studies Have students check their local newspapers and magazines and report back on the kinds of things people are doing about recycling and pollution.

Science Have students investigate the following questions. Are there really robots? If so, what kinds of things do they actually do?

Dear Students,

I am Roberta Recycle, Queen of Recycling Robots. We are special robots made out of recycled materials. Our job is to remind everyone to recycle. I come to you for help. We all know that pollution and too much trash are problems. But we can do something about it. We can let people know about recycling.

Please help me spread the word about recycling by creating a special helper – your own Recycling Robot. A Recycling Robot is a robot made out of juice boxes, tape, and recycled and scrap materials like bottle caps, straws, and yarn.

To make a Recycling Robot, there are certain rules you must follow:

1. It must be built so that it is very sturdy and can be moved easily from place to place.
2. It has to hold a sign with a message to remind people to recycle.
3. It must be built of Robo-Blocks. A Robo-Block is 10 juice boxes put together with tape to make a strong, sturdy block of any shape.
4. A Recycling Robot must be made of at least 10 Robo-Blocks. You may use more.
5. A Recycling Robot must have a name.

We would like each of you to make a Robo-Block from 10 juice boxes. Then, on the Robot Recording Sheet, draw a picture of the Robo-Block you made. Write a number sentence that matches your Robo-Block and answer the three questions. Give this to your teacher as a record of your work. Finally, combine all the Robo-Blocks in your group to create a Recycling Robot.

Good luck and happy recycling!
Thanks for your help,

ROBERTA RECYCLE,

Queen of the Recycling Robots

Robot Recording Sheet

My name: _____

Sample Robo-Block Picture:

Sample Number Sentence(s)

$5 + 5 = 10$

$2 + 2 + 2 + 2 + 2 = 10$

$5 \times 2 = 10$

$2 \times 5 = 10$

My Robo-Block Picture:

My Number Sentence(s)

How many <u>juice boxes</u> did your group use altogether? _____

How many <u>Robo-Blocks</u> did your group use altogether? _____

What did you name your robot? _____

Fold

THANK YOU

Dear Wonderful Students,

You really are wonderful! I knew I could count on you. Your Recycling Robots are fantastic. They will really help to spread the word about recycling!

You are helping to make the world a cleaner, healthier, and happier place to live by having your Recycling Robots send out the message.

Thank you so much,

ROBERTA RECYCLE

Queen of Recycling Robots

Dear Boys and Girls,

Hi! It's me, Jack. The Jack from the famous story, *Jack and the Beanstalk*. I am writing to ask for your help with a problem. I noticed that the giant who lives at the top of the beanstalk was missing one of his shoes. I figure that if I could make him a new shoe, I could get on his good side. The only thing I was able to find out was that his foot is as long as the feet of 6 first-graders put together end to end.

Now here are the two things you can help me with.

1. Can you tell me how long six of your feet are when placed end to end?

2. Can you make me a picture of the giant's shoe that's the right size?

 I'll be back in a few days to get your written answers. Just leave them on the teacher's desk, and I'll find them. Please make sure to leave your pictures of the shoe in a place where I'll be able to see them.

Thanks for your help!

Your pal,

Jack

(from the Beanstalk)

THE GIANT'S SHOE

Grade Level: 1

This problem can be made appropriate for students in Grade 2 as well. See Customizing the Project, page 47.

Overview

"Jack" from the classic fairy tale, *Jack and the Beanstalk*, asks the class for help. It seems that the giant has lost one of his shoes. Jack would like to make a new shoe for the giant, in hopes that this will get him on the giant's good side. Jack knows that the giant's foot is the length of 6 first-graders' feet placed end to end. To help Jack out, students measure, record, add, and use this information to create a full-sized picture of a shoe for the giant.

Objectives

- To use a group problem-solving approach to collect, organize, and utilize data in order to produce a physical product
- To select and use measuring tools appropriately

The Mathematics

- *addition*
- *subtraction*
- *counting*
- *data collection*
- *data analysis*
- *data organization*
- *measurement (standard units and nonstandard units)*

Time Needed   four to five 40-minute periods

Grouping   groups of four to six students

Preparation and Materials

Day 1

- Any version of the book, *Jack and the Beanstalk*, 1 teacher copy
- *The Giant's Shoe* story, 1 teacher copy, page 49
- Jack's letter, 1 per student, page 50 (Instead of, or in addition to, having copies on letter-sized paper, you may want to copy the letter onto chart paper to make the letter easier for all to read together.)
- An envelope "made from" paper beanstalk leaves or decorated with beanstalks to hold the letters (If you choose to use a chart letter only, you can roll up the chart and put it in a beanstalk-decorated mailing tube or simply roll it up and secure it with a rubber band.)

Days 2, 3, and 4

- The book, *How Big is a Foot?*, by Rolf Myller, (optional) 1 teacher copy
- Individual Recording sheets (optional), 1 per student, page 51
- Group Recording sheets (optional), 1 per student, page 52
- Drawing paper, 1 sheet per student

- White or light-colored roll paper, 1 six-foot length per group
- Rulers, 1 per student
- Tape measures (optional), 1 per group
- Yardsticks, 1 per group
- Yarn (optional), 1 six-foot length per group
- Nonstandard measuring manipulatives (such as Unifix Cubes, links, Snap™ Cubes, paper clips), as needed
- Crayons, 1 set per student
- Markers, 1 set per group
- Scissors, 1 per group
- Tape, 1 roll per group

Day 5

Thank-you notes from Jack, 1 per student, page 53

PROJECT: THE GIANT'S SHOE

Day 1: Special Delivery

Deliver

You may want to tell students that this is a new version of the story.

Read aloud to the class any traditional version of *Jack and the Beanstalk* up to the point when Jack arrives at the top of the beanstalk for the first time. Stop reading from the book and continue reading *The Giant's Shoe* story. Discuss the giant's problem and Jack's reason for wanting to help.

Then "discover" Jack's letter. The envelope (or rolled-up chart letter) might be delivered to the door or "found" in your room. Distribute the letters (and/or display the chart letter).

Read

Give time for students to read the letter silently. Then have them take turns reading aloud or read it aloud to them. Discuss vocabulary and terminology together.

Discuss

The problem is woven into the context of the letter. To help students interpret the letter and understand the task, ask

1. So, what's the problem?

Use some or all of the following questions to guide the discussion and analysis of the letter.

Who wrote the letter? (Jack)

In the letter, it says that the giant has a problem. What is it? (The giant is missing a shoe.)

What does Jack want us to do? (Two things: Tell him how long 6 first-graders' feet are altogether and make a picture of the shoe in the right size)

Why? (So the giant will like Jack)

Let's look at our shoes. What is the same about them? What is different? Do you know your shoe size? (Allow time for students to discuss and share information in order to build a connection to previous knowledge and make the problem more motivating.)

What do we know about the missing shoe? (It is as long as 6 of our feet put end to end.)

Once students are clear on the information given in Jack's letter, ask

2. What materials do you think you will need to solve the problem?

Depending on the class, it might be a good idea at this time to direct students to think about the need for measurement in the problem. This will help them to think about the measurement tools they may want to use.

List class responses on the board or on a chart. Have students justify why they feel they need a certain material. As measurement materials are suggested by students, be open to listing both standard and nonstandard measurement tools. Tell students you will gather necessary materials and have them ready for work on the next day.

Day 2: Let's Get to Work!

Have all materials ready for students. Form work groups.

Re-read the letter from Jack either from the chart or student copies of the letter. Begin with a discussion reviewing the problem so everyone is reminded of what the task is.

It may be appropriate here to do a mini-lesson on measuring with some of the tools suggested by students. Reinforce the ideas of starting at the edge of the measuring tool, measuring straight across the object to the very opposite side, making the edges of the measuring tools touch each other with no spaces in between, and measuring more than once to be sure.

Ask students to save all their work for you including all recordings of measurement and addition of data. Have students begin work. Circulate, assist, and observe.

> Six students is ideal for a group, but smaller groups can still complete the problem by using strategies like measuring a group member's foot more than once.

Common Student Approaches

At this point, students have the choice of using any measuring tools they wish. Many students will tend to work very independently since they do not need group cooperation when measuring their own feet; as a result, some groups will have a variety of units of measure. Other groups may organize themselves so that all group members use the same measuring tools and their results are in the same unit of measure.

Once students have completed their individual foot measurement, it is interesting to see what they do next. Some groups will begin to add up their data. Difficulties occur when they have used different units of measure. This is a great opportunity to discuss the need for standardized measurement in our world. This may occur today on Day 2, or it may be brought up during the discussion at the beginning of Day 3.

> One option is to have students use the individual and group recording sheets provided. Consider though that if you supply these, you will be directing students toward a method of organizing their data instead of having them find their own methods.

Teacher Role

While students are working, circulate, assist, and observe. In addition, use this time to begin assessment. Keep these assessment questions in mind:

What math are students using?

What organizational ideas are students using?

What tools are students using?

What measurement techniques are students using?

How accurate is the measurement?

What are students doing well?

What difficulties are students having?

Day 3: Time to Share

Begin with a whole class discussion to give students the opportunity to think about their partial solutions in light of what their peers share. Have all students sit in a circle with their work. Ask the following questions and discuss answers together:

What did you do first? Why? Was it a good choice?

What has worked well for you?

What problems have you had? Who else has had these same problems? What have you done about these problems?

Who knows how long their foot is?

How did you find out? Can you show us how you did it?

How are you keeping track of the foot measurements in your group?

Review the written work completed so far. Have students report their foot length in terms of the unit of measure they used. For example, they may report "My foot is 9 Unifix Cubes long," or "My foot is eight inches long."

If students in the class have used different units of measure to determine their foot lengths, discuss the problem of how to compare the lengths.

Summarize the discussion with these questions:

Can we compare fairly and easily using different units of measure? (No)

What do you need to do with your group's foot measurements? (Decide on one standard unit of measure for all feet)

Ask for suggestions as to the best standard unit to use to measure a person's foot. Some students may recommend using tools like Unifix Cubes. This is fine for the purposes of the project; this response provides an opportunity to discuss the possible advantages of using a more widely recognized measuring tool, both in this problem and in other situations. (For example, ask whether a clerk in a store would know how much wool or yarn to give a customer if the customer asked for it measured in paper clips or Unifix Cubes.) Talk about rulers and their valuable role as a standardized measuring tool for inches. Do a mini-lesson on using rulers if students have had little or no prior experience.

Have students return to their groups to resume work. Again, as you did on Day 2, circulate, observe, assist, and assess.

An Optional Activity

If you have time, you may want to list all individual foot measurement responses on the board or chart paper. Discuss the list and ask students to note the range of foot sizes in the class. This will be difficult if responses are in different units of measure. Ask students to measure again using a ruler and record the information in inches on

> At this point, you may wish to read aloud *How Big is a Foot?* which describes a situation in which there are different units of measure, making comparisons difficult.

> The Optional Activity might be appropriate for Day 3 or for Day 4.

a recording sheet or other paper. Students who have measured with a ruler the first time around may check their work by measuring again or assist other students by checking their measurements.

Day 4: Still Working

Begin with a whole class share session. Ask the following questions and discuss:

What has worked well for you?

What problems have you had? How are you trying to solve these problems?

How many of you have started to make the shoe?

How do you know when you are ready to make the shoe?

Some students may have difficulty organizing their data to combine results. Ask the class as a whole to share the methods their groups used. Have students facing difficulty try out one of the methods discussed for combining results.

As students begin making the shoe pictures, remind them that they have determined only the length of the shoe. The height is up to them. The most appropriate way for students to make the shoe is to create a flat side view of the shoe on roll paper. Some may want to create a 3-D shoe, but this will take more time. If you have time and want to go this route, enjoy! In any case, encourage students to use whatever colors and styles they think the giant would like.

> Use the recording sheets if you think they will be helpful here.

Day 5: Closure

Have students share their written responses with the class and discuss. Use these questions to guide the discussion:

Remember that Jack said the giant's foot was "about" the length of 6 first-graders' feet. What does "about" mean? (close to but not exactly the same)

Looking at the different groups' answers, are they about the same?

Why are they at all different?

Leave the written answers on your desk for Jack to find when he stops by. Leave the shoe pictures on display so Jack can see them. While students are out of the classroom, arrange for Jack's thank-you note to arrive. Read the note aloud with students.

CUSTOMIZING THE PROJECT

Supplying students with the individual and group recording sheets gives them more direction and organization. A more challenging approach is to not use these support pieces and have students create their own ways of recording and organizing information.

Another interesting and more challenging option is to change the letter requirements so that the giant's shoe size is the same as the length of all the students' shoes placed end to end. This will move groups to work together to determine a standardized measure for the whole class and a system of adding lots of numbers to arrive at a significantly larger answer. This option calls for a great deal of cooperation between groups both in the math and in the creation of the actual shoe.

ASSESSMENT

For additional tips on assessment, see Appendix 1, Assessment, pages 97-99.

Use questions like these as you consider the final product and the process you observed over the course of the project:

What measuring tools were used? Was measurement accurate?

How was cooperation within the group?

What kind of math was used? Was addition accurate?

How was data organized and used?

Math Journal Writing or Interviews

Ask students to answer one or more of the following questions either in writing or orally:

What mathematics did you use to solve the problem?

How can you be sure you are measuring accurately?

Why are standardized measures like inches helpful?

Why are rulers helpful math tools?

What is a good strategy for combining numbers?

What is a good strategy for keeping numbers organized?

If you were to share one strategy or method that worked particularly well for you in solving this problem, what would it be?

CURRICULUM CONNECTIONS

Literature As part of the project, read *How Big is a Foot?* by Rolf Myller. There are also a great many versions of *Jack and the Beanstalk*, including a funny, colorful one by James Marshall and a more scary, graphic one by Steven Kellogg.

Math Have students look at actual shoes and shoe sizes. Ask what the shoe sizes mean. Contact a shoe store to have a representative come in and talk about measuring feet and how math is used in this process.

Art Ask students to create sketches of other styles of shoes for the giant in smaller scale. As a starting point for their sketches, direct students to look at the shoes of their classmates and note the variety of color, style, cut, shape, and material.

Music Ask students if they can find or make music that reminds them of a big, loud giant. Distribute rhythm instruments and have the class play them and march like giants.

Language Arts Giants are fascinating characters to children. Have students pretend they are Jack's giant and write a letter telling what it is like to be a giant.

Social Studies Different countries around the world use different types of measurement, among them the metric system and standard U. S. (or customary) system. Ask students to find examples of different kinds of measures at home in the kitchen or garage.

Science Bean plants are among the easiest plants to grow in a classroom and also can get quite long. Have students plant bean plants and then measure and record their growth over a set period of time.

(Note to teachers: Read aloud any version of Jack and the Beanstalk *to the point in the story when Jack reaches the top of the beanstalk for the first time. Then continue to read aloud the following.)*

The Giant's Shoe
(or Jack and the Beanstalk, continued)

When Jack finally reached the top of the beanstalk, he was amazed at what he saw. Everything was larger than he could imagine. Jack saw a house that was so big that it had to be the home of a giant. The giant's mailbox was three times as tall as Jack. The giant's front door was as tall as Jack's whole house.

Jack slipped into the giant's house, found a good hiding place, and waited to see what the giant looked like. When Jack finally got to see the giant, he heard the giant complaining that he was missing one of his giant shoes. This made the giant very angry because he had only one pair of shoes. Jack thought it would make the giant happy to have a new shoe, and so Jack decided to make one for the giant. Jack hoped that this would get the giant to like him.

In order to be able to make the giant's shoe, Jack needed to find out the giant's foot size. He couldn't ask the giant for his size because the shoe was going to be a surprise. He couldn't measure the giant's foot because the giant might catch him and eat him. After speaking with the giant's housekeeper, Jack learned that the giant's foot was just about as long as 6 first-graders' feet put together.

(Note to teachers: Now read Jack's letter to students.)

Dear Boys and Girls,

Hi! It's me, Jack. The Jack from the famous story, *Jack and the Beanstalk*. I am writing to ask for your help with a problem. I noticed that the giant who lives at the top of the beanstalk was missing one of his shoes. I figure that if I could make him a new shoe, I could get on his good side. The only thing I was able to find out was that his foot is as long as the feet of 6 first-graders put together end to end.

Now here are the two things you can help me with.

1. Can you tell me how long six of your feet are when placed end to end?

2. Can you make me a picture of the giant's shoe that's the right size?

I'll be back in a few days to get your written answers. Just leave them on the teacher's desk, and I'll find them. Please make sure to leave your pictures of the shoe in a place where I'll be able to see them.

Thanks for your help!

Your pal,

Jack

(from the Beanstalk)

Individual Recording Sheet

My name: _____

How long is your foot?

    Trace your foot on the back of this paper.

    Measure it.

What did you measure with? _____

_____

_____

How long is your foot?

My foot is _____ **long.**

Group Recording Sheet

My name: _____

Group members: _____

_____

_____

_____

My foot is _____ **long.**

Six of our feet together are _____ **long.**

The giant's shoe needs to be _____ **long.**

Fold

Dear Students,

Thanks for the great job you did in making another shoe for the giant. I'm his best friend now, thanks to you. You really know how to work together to solve a problem.

The giant was so happy when he saw that new shoe. He told me to give you a big

THANK YOU!

Bye and thanks!

Your friend,

A Special Invitation To You

Dear Students,

You are cordially invited to shop at The Necklace Shop. Your teacher has been kind enough to set up our shop in your class. Our shop sells special "Make-Your-Own Necklaces." Because you are new customers and we want to keep your business, we would like to give you 97 cents to buy all the materials you need to make your necklace.

Here are the prices of the materials we have for sale.

<u>Blue Beads</u> cost 5 cents each.

<u>Red Beads</u> cost 2 cents each.

<u>White Beads</u> cost 1 cent each.

<u>1 inch of yarn</u> costs 1 cent.

Here are the steps you should follow to take advantage of our make-your-own-necklace invitation.

1. Experiment with bead patterns until you find one you like. Show your necklace pattern and plan by drawing its picture on The Necklace Shop Planning Form.

2. Use the information on your Planning Form to fill in The Necklace Shop Materials Order Form. You must spend exactly 97 cents – no more and no less.

3. Show your completed Planning Form and Materials Order Form to your teacher who will check and approve your work.

4. Bring your 97 cents, Planning Form, and Materials Order Form to the Necklace Shop clerks. They will serve you with a smile!

5. Have fun making your necklace!

Sincerely,
The Necklace Shop Owners

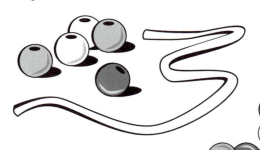

THE NECKLACE SHOP

Grade Level: 2

Overview

Students receive an invitation to shop at The Necklace Shop, a "make-your-own-necklace" specialty shop set up in the classroom. Students create a plan and fill in an order form for a necklace that costs exactly ninety-seven cents. When they are ready with a completed, approved plan and order form, students receive their beads and yarn and make their own necklaces.

Objectives

- To control two linked variables (money and a pattern) while solving a complex problem
- To use a diagram as a component of an overall problem-solving strategy
- To use mathematics to create a real, physical object

The Mathematics

- *addition*
- *money*
- *calculator use (optional)*

- *repeated addition*
- *patterning*
- *measurement*

- *skip counting*
- *multiplication*
- *controlling multiple variables*

Time Needed   four to five 40-minute periods

Grouping   Students should work independently.

Preparation and Materials

Day 1

- Necklace Shop letter, Planning Form, and Materials Order Form, 1 each per student, pages 62-64.
- Large envelope in which to deliver the letters and forms (You may want to decorate the envelope with magazine pictures of necklaces.)
- The Necklace Shop (This can consist of a desk or two with boxes for the beads, rolls of yarn, scissors, a yardstick or ruler, and a sign.)
- Necklace Shop supplies (Assuming that students use several of each color of bead, a class of twenty-five students will require about 15 yards of yarn and at least 700 white beads, 300 red beads, and 150 blue beads. Instead of using actual beads, you may wish to dye pieces of macaroni in the appropriate colors.)
- Pre-packed envelopes or packets each containing 97 cents in real or play coins, 1 per student
- Manipulatives that come in at least three colors such as counting bears or Unifix Cubes

- Sample necklace created from materials in the shop (optional)

Days 2, 3, and 4

- Play coins or real coins (for trading in)
- Masking tape (for wrapping the ends of the yarn to function like shoelace tips to make stringing the beads easier)
- Calculators (optional)
- Crayons in colors to match the beads, 1 set per student
- Drawing paper
- Measuring tapes

Days 3, 4, and 5

Arrange for "staffing" of the Necklace Shop. It will be very difficult for you as the teacher to be the shop clerk and still work around the room with students. Consider having volunteer parents come in to staff the shop. Their task is simple: take the students' money and approved forms, help count out the correct number of beads, and cut the correct amount of yarn. Other options are to invite former students from upper grades to be clerks or have students who finish early take turns staffing the shop.

Day 5

Camera for photographing completed necklaces (optional)

PROJECT: THE NECKLACE SHOP

Day 1: Special Delivery

Deliver

Have the envelope with Necklace Shop letters, Planning Forms, and Materials Order Forms delivered to your room. It can be fun to display or wear a sample necklace made from materials from the shop. This also stimulates a lot of interest. Open the envelope for the class and pass out the letters.

Read

Give time for students to read the letter silently. Then have them take turns reading aloud. Discuss vocabulary and terminology together.

Discuss

The problem is woven into the context of the letter. To help students interpret the letter and understand the task, ask

1. So, what's the problem?

Use some or all of the following questions to guide the discussion and analysis of the letter.

Who wrote the letter? (The Necklace Shop owners)

What have you been invited to do? (Shop at The Necklace Shop and make our own necklaces)

Who has ever worn or seen necklaces? Can you describe them? (Allow time for students to discuss what they know about necklaces to build a connection to previous knowledge and make the problem more motivating.)

At this point, take out the materials and set up The Necklace Shop, or, if you have already set up the shop, call students' attention to it. Show the different colored beads and the yarn to make the task more concrete for students.

What else came with the letter? (A Planning Form and a Materials Order Form)

Let's look at the Necklace Planning Form. What would make a good plan for a necklace? What would it look like? (Clear, easy to read, uses colors corresponding to real beads, shows a pattern clearly)

Let's look at the Materials Order Form. What do you have to do with this? (Fill in all the blanks. Tell how many of each bead we will buy and how much those beads will cost. Tell how much yarn we will buy and how much the yarn will cost. Tell what the total of all our purchases is.)

What does the letter say you need in order to go to the shop and buy things? (A completed and approved Planning Form, Materials Order Form, and the required money)

How about money? What does the letter say about money? (It says we must spend exactly ninety-seven cents.)

Now is a good time to give out envelopes or packets containing the pre-counted coins or play money. Have students count and verify that their envelopes contain ninety-seven cents. The combination of coins you use to make the ninety-seven cents is something to think about. Whatever combination of coins you use, have a class set of coins available on Days 2, 3, and 4 so that students can trade in coins as they work. For instance, if students have a dime and want to buy one-cent beads, they may want to work with pennies and trade in that dime for ten pennies. At the end of this session, collect the envelopes for redistribution on Day 2.

Speaking of money, do the beads all cost the same? (No, the blue beads cost 5 cents each, red beads cost 2 cents each, and white beads cost 1 cent each.)

Do you have to pay for anything else? (Yes, yarn costs 1 cent for each inch.)

It says in the letter that you need to have a pattern for your necklace. What is a pattern? (An arrangement that repeats itself in some way. Students may answer this by demonstrating a clapping pattern or using manipulatives to build one.)

What types of patterns are there? (Students can demonstrate a variety of patterns, again either by clapping or with manipulatives. Urge them to show more complex pattern ideas. More complex patterns make the problem more challenging.)

Okay, so you need to create a pattern plan, then fill in an order form, get your money together, and then go shopping. Anything else? (Open question time)

Once students have finished discussing the information given in the letter, ask

2. What materials do you think you will need to solve the problem?

List class responses on the board or on a chart. Have students justify why they feel they need a certain material. Tell students you will gather necessary materials and have them ready for work the next day.

Day 2: Let's Get to Work!

Have all materials ready for students. Briefly review the questions and information from Day 1 so everyone is clear on the task and parameters. Have students begin work. Circulate, assist, and observe.

Common Student Approaches

Students begin in many different ways. Some will start by trying to figure out how much yarn is needed. This is a good, simple, first step. Others will begin drawing the plan or using manipulatives to physically plan out their necklace designs.

When students figure out how much yarn they need, they may try to measure their necks or heads. Some will try with a ruler, which is a tough trick but can yield an approximate answer. Others will use yarn or string, measure around the neck or head, then measure the piece of yarn with a ruler. This may be a nice opportunity for students to learn how useful a tape measure can be. After a few goes at trying to measure their necks and heads with a ruler, students see the measuring tape as an absolutely brilliant invention!

A key part of this problem is the experience with two linked variables: money and a pattern. As the necklace pattern is created, money is spent. Both must be controlled to successfully meet all the requirements of the problem. As students work through the Materials Order Form, there are opportunities for them to use repeated addition and multiplication. Some students will keep a written running record. Some students may try to keep all this information on their calculators.

To use or not to use calculators? That is the question. If you want students to practice and apply double-digit addition of multiple addends with regrouping and experience how a large addition problem can be broken up and done a bit at a time, you may *not* want them to use calculators. If these concepts are not primary objectives, this problem may be a good opportunity for showing students that a calculator is an appropriate tool for keeping track of totals and/or checking addition.

Teacher Role

While students are working, circulate, assist, and observe. Keep these assessment questions in mind:

What math are students using?

What organizational ideas are students using?

What tools are students using?

What are students doing well?

What difficulties are students having?

Day 3: Time to Share

Begin with a whole class discussion to give students the opportunity to think about their partial solutions in light of what their peers share. Have groups sit in a circle with their work. Ask the following questions and discuss answers together:

What did you do first? Why? Was it a good choice?

What has worked well for you?

What problems have you had? Who else has had these same problems? What have you done about these problems?

Who knows how much yarn they need? How did you figure this out? (Caution students to plan for enough yarn so that the necklace will slip over their heads.)

Who knows how much money they have spent so far? How do you know? Why is it important to know?

How are you keeping track of what you are doing? How do you know how much money has been spent or how many beads have been used?

This is a good time to remind students that the information on the Planning Form and Order Form must match. You may want to do a short example on the board for a pattern of three or four beads to illustrate.

What are some of the patterns we have so far? (Let students share pattern ideas.)

After the discussion, direct students to continue working. Again, as you did yesterday, circulate, assist, observe, and assess.

It is possible that some students will be ready to shop on this day, so the shop should be staffed and ready. Remember to check and approve students' work before you let them go shopping. As students finish shopping and make their necklaces, one option is to have them help as clerks at the shop or have them try one of the Curriculum Connection suggestions on pages 60-61.

Day 4: Still Working

Have the Necklace Shop set up, staffed, and ready for business again. Begin with a whole class share session. Have students share any completed necklaces. Ask the following questions and discuss:

What has worked well for you?

What problems have you had?

How are you trying to solve these problems?

Before students continue their independent work, remind them that complete necklace Planning Forms and Materials Order Forms must be checked and approved by you before they can go to the Shop.

Day 5: Closure

Ask all students to wear their necklaces. Take a group photo. Students are proud of their creations and will often wear them afterwards throughout the year. Use these questions to help summarize the week's work:

Was there was any one right way to make the necklace? Why or why not?

What are some of the ways you made your necklaces?

Describe some of the necklace patterns here. How are they the same? How are they different?

You may want to discuss whether a more complex pattern always results in a more beautiful necklace. Students may be surprised to discover that they find more aesthetic enjoyment in necklaces made from simpler patterns.

CUSTOMIZING THE PROJECT

Depending on your class or individuals within the class, you may want to change the money parameter. You could simplify the problem by having students spend within a range; for example, you can tell them that they must spend between 80 cents and a dollar. Or, you may want to lower the total amount of money to 50 cents. You can also increase or decrease the complexity of the problem by offering a wider or smaller variety of beads at different prices. Older students may not need the coin manipulatives.

ASSESSMENT

For additional tips on assessment see Appendix 1, Assessment, pages 97-99.

Review completed necklaces, necklace plans, and order forms. Think about these questions as you consider the final product and the process you observed over the course of the project:

Does the necklace have a pattern? How complex is the pattern?

Does the pattern in the necklace match the pattern on the necklace plan?

Is the necklace plan clear and easy to understand?

Is the order form mathematically accurate?

What kind of math (counting, skip counting, column addition, multiplication) was used to complete the order form?

Does the order form match the necklace plan?

Was data well organized?

To what degree did the student participate in the process? How much assistance was needed from the teacher?

Math Journal Writing or Interviews

Ask students to answer one or more of the following questions either in writing or orally:

What mathematics did you use to solve the problem and complete your necklace?

How did you keep track of the money, the beads, and the yarn in your plan?

Is there a good strategy for adding up numbers that are the same, like a 5 and a 5 and a 5?

If you were to share one strategy or method that worked particularly well for you in solving this problem, what would it be?

CURRICULUM CONNECTIONS

Literature Read aloud *Piggins*, by Jane Yolen. Piggins the butler solves the mystery of Mrs. Reynard's missing necklace. This is a great, fun-to-read, picture book.

Math Ask students to create complex patterns with numbers and challenge their fellow students and/or the teacher to figure them out. The patterns might involve concepts such as odd and even, double digit and single digit, skip counting, and backwards and forwards counting. Remind students that they must know the answers to the challenges they pose.

Art Invite students to make necklaces out of other materials such as baked clay, metals, and recycled materials.

Music Have students look for patterns in songs and printed music. Ask if they can find examples in music of repeated patterns both in what they hear and in the notes they see.

Language Arts Ask students to write an imaginative story about their necklaces that includes the answers to some or all of these questions: Where was it from? Whom did it belong to originally? Is there anything special we should know about it? Does it have an intriguing history behind it?

Social Studies Have students research the kinds of jewelry worn in different countries, both ancient and modern. Find photos or pictures of the jewelry. Discuss how the necklaces are similar to the ones students made, what patterns they have, and what sort of gemstones are used.

Science Explain that many famous necklaces feature beautiful and valuable gems and gemstones. Then have students look for and discuss the answers to these questions: What makes a gem or gemstone valuable or not? What special qualities do they have? Where can they be found?

A Special Invitation To You

Dear Students,

You are cordially invited to shop at The Necklace Shop. Your teacher has been kind enough to set up our shop in your class. Our shop sells special "Make-Your-Own Necklaces." Because you are new customers and we want to keep your business, we would like to give you 97 cents to buy all the materials you need to make your necklace.

Here are the prices of the materials we have for sale.

<u>Blue Beads</u> cost 5 cents each.

<u>Red Beads</u> cost 2 cents each.

<u>White Beads</u> cost 1 cent each.

<u>1 inch of yarn</u> costs 1 cent.

Here are the steps you should follow to take advantage of our make-your-own-necklace invitation.

1. Experiment with bead patterns until you find one you like. Show your necklace pattern and plan by drawing its picture on The Necklace Shop Planning Form.

2. Use the information on your Planning Form to fill in The Necklace Shop Materials Order Form. You must spend exactly 97 cents – no more and no less.

3. Show your completed Planning Form and Materials Order Form to your teacher who will check and approve your work.

4. Bring your 97 cents, Planning Form, and Materials Order Form to the Necklace Shop clerks. They will serve you with a smile!

5. Have fun making your necklace!

Sincerely,

The Necklace Shop Owners

Necklace Shop Planning Form

Draw a picture of your necklace plan here.

Necklace Shop Materials Order Form

Name _____

Please fill in all the blanks on this form.

Yarn costs 1 cent for 1 inch.

How many inches? _____   How many cents? _____

Blue Beads cost 5 cents each.

How many Blue Beads? _____   How many cents? _____

Red Beads cost 2 cents each.

How many Red Beads? _____   How many cents? _____

White Beads cost 1 cent each.

How many White Beads? _____   How many cents? _____

Total - How many cents? _____

THE NECKLACE SHOP

PARK RANGER RALPH
NEW JERSEY PINE BARRENS
U.S.A.

Dear Students,

While patrolling recently in a cranberry bog deep in the New Jersey Pine Barrens, I came upon this mysterious, sealed box, sunk in the bog.

I have measured the box with great accuracy. It is 12 inches wide, 16 inches long, and 2 inches high.

I believe there may be money inside. In fact, I am almost positive it belonged to the notorious Two-Dollar Pete, a bank robber who had the strange habit of stealing only two-dollar bills.

I would like you to open the box and find out what is inside. But first, because I love math as much as you do, I'd like to challenge you to figure out the answers to these questions:

1. How much money do you think Pete could fit into the box?
2. If you shared that amount of money with the members of your class, how much would each person get?

No matter what your answers are, be sure to show clearly how you got them.

Good Luck,
Park Ranger Ralph

P.S. Once you figure out the problem, you may open the box. Whatever is inside is yours to keep with my compliments.

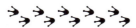 

Two-Dollar Pete

Grade Level: 2

This project can be interesting for third and fourth graders as well. See Customizing the Project, page 71.

Overview

An old box? An old, sealed box? An old, sealed box with treasure inside? A strange, well-worn box arrives in your classroom with a note from Park Ranger Ralph, the math-loving ranger who found the box. He believes it belonged to the notorious Two-Dollar Pete, a bank robber who had the strange habit of stealing only two-dollar bills. But the box can't be opened until the class figures out how much money could actually be inside and how the money could be divided among the members of the class!

Objectives

- To use spatial reasoning, measurement, and calculation to solve a complex problem
- To develop a mathematical strategy to determine the amount of money that might fit in a box with given dimensions

The Mathematics

- *measurement*
- *capacity*
- *repeated addition*
- *double-digit addition*
- *place value*
- *multiplication*
- *spatial reasoning*
- *division*

Time Needed   four to five 40-minute periods

Grouping   This could be an individual problem if there are enough materials, or it can be done by partners or small groups of up to four.

Preparation and Materials

Day 1

- Letter from Park Ranger Ralph, 1 per student, page 73
- Large envelope addressed to the class to hold the letters
- One box, 12 in. by 16 in. by 2 in. (This is a standard-size clothing box. Any size box will do; just make sure to change the dimensions in the letter to approximate those of the box you use.)
- Contents of the box
    - A reward for the class. One possibility is foil-wrapped chocolates that resemble coins. Another is candy bars that have brand names that are amounts of money. Wrap the reward in newspaper to give the box some weight and to prevent the rewards from rattling.
    - Pete's poem, 1 copy, page 74
- Paper for wrapping the box. Use old brown paper or the like – the dirtier and more scuffed the better. Scrawl messages on the paper such as "Beware!", "Don't Touch!", and "Property of Pete."

Days 2, 3, and 4

- Calculators, at least 1 per group
- Rulers, at least 1 per group
- Play money bills, at least a class set
- Several real dollar bills (optional)
- Copy paper, 1 ream
- Scissors, 1 per group
- Cardboard or postboard

PROJECT: TWO-DOLLAR PETE

Day 1: Special Delivery

Deliver

Attach the envelope with the letters from Park Ranger Ralph to the wrapped box and have it delivered to your room. Walk around it cautiously. Put it in the center of the room or on a table. Invite students to come closer, but not to touch. Someone will probably notice the envelope addressed to the class and request that it be opened. Do so and pass out the letters.

Read

Give time for students to read the letter silently. Then have them take turns reading aloud. Discuss vocabulary and terminology that might be new to students to ensure comprehension.

Discuss

All of the information needed to solve the problem is in the letter. It is not even necessary for students to see or touch the box although its presence is a real motivator. During discussion it is important that students get a good understanding of just what the problem is and what they are required to do. To help students interpret the letter and their task, ask

1. So, what's the problem?

Use some or all of the following questions to focus the discussion and analysis of the letter.

Who wrote you the letter? (Park Ranger Ralph)

What does the Park Ranger think is in the box? (money)

What was unusual about the bank robber mentioned in the letter? (He stole only two-dollar bills.)

What does the Park Ranger want us to do? (Two things: figure out how much money could be in the box and how much of the money each person in the class should get)

Why does he want us to do that? (He loves math.)

An excellent bit of homework for students is to find out if there actually are such things as two-dollar bills and if anyone can bring one in.

What happens after we figure out how much money is in the box and how much each person should get? (We get to open the box and keep whatever is inside.)

Do you need to measure the box? (No, the measurements are given in the letter.)

According to the letter, what size is the box? (12 in. wide, 16 in. long, 2 in. high)

Once students are clear on the information given in the letter, ask

2. What materials do you think you will need to solve this problem?

List responses on the board. Ask students why there is a need for these items and how they will be used. This gets students to begin thinking about strategies and approaches. Tell students you will have the necessary materials ready for the next session, and work will begin then.

There may be requests for lots of real money. Explain to the groups that you do not have lots of real money. (You are, after all, a teacher.) You may wish to offer to bring in a few crisp dollar bills for each group to use, but explain that groups will have to develop their own alternatives (like cutting same-size bills out of copy paper or using play money).

Day 2: Let's Get to Work!

Have all materials ready for students. Form student groups. Briefly review the questions from Day 1 so everyone is clear on the task and the parameters. Then ask

Are there any questions or concerns about the problem?

Do you all know what you are required to do?

Explain to students that this problem requires a good deal of group discussion and cooperation because they must together determine one method for solving the problem. Remind students that they need to be able to justify their answers and show how they arrived at them. Once you are satisfied that students have a satisfactory understanding, have them begin work.

Common Student Approaches

Students may flounder at first and have no idea where to begin. Give them time. Remind them to look at the letter and think about what information is given. As keys to solving the problem, help students to focus on the dimensions of the box and the fact that there are supposedly two-dollar bills in it. Try to help them to "think measurement."

Once they get going, students approach this problem in a variety of ways. Some groups will actually build a box replicating Pete's box. Some will recreate only the bottom of the box. Some will make one 2-inch stack of bills; others will try to make stacks and stacks of bills to fill the prototype box. There is fertile ground for later discussion here, particularly about which approaches are most efficient and why.

Another interesting approach students may try is weighing the box and then weighing stacks of bills. This is good thinking except that it becomes difficult to factor in the weight of the actual cardboard box and the wrapping. Another problem is that there is only one box and many groups. If you want to avoid these difficulties, you can make the simple rule that no one may touch the box. All information must come from the letter.

Remind students that they will need to show their work, so they should keep all calculations and papers. It is also helpful to remind them of the importance of attaching labels to numbers (for example, 24 inches, 24 dollars, 24 bills) to avoid confusion.

Teacher Role

While students are working, circulate, assist, and observe. In addition, use this time to begin assessment. Keep these assessment questions in mind:

What math are students using?

What strategies are students developing?

How is the group cooperating and dividing up tasks?

What are students doing well?

What difficulties are students having?

What tools are students using?

Day 3: Time to Share

Begin with a whole class discussion to give students the opportunity to share and evaluate their partial solutions. Have groups sit in a circle with their work. Ask the following questions and discuss answers together.

What did you do first? Why? Was it a good choice?

What has worked well for you?

What problems have you had?

What is your plan for how to figure this out?

Do you need lots of bills to figure this out? Why or why not?

Is it important that Pete stole only two-dollar bills? Why?

As students respond, take advantage of the opportunity to reinforce valuable concepts or ideas that students suggest. After discussion and shared answers, have students begin work. Again, as you did yesterday, circulate, assist, observe, and assess.

Day 4: Still Working

Begin with a whole class share session. Ask the following questions:

What has worked well for you?

What problems have you had?

How are you trying to solve these problems?

Tell groups that once they arrive at an answer they should prepare a presentation for the next class session that explains what they have done and why. Following any discussion, have students return to work. Again, circulate, observe, assess, and when necessary, assist.

Day 5: Closure

Have groups present their solutions. Suggest that they tell about their problem-solving process step by step, maybe even using the actual materials, and explain how they finally came to their answers. Discuss why there may have been different answers and approaches.

Clearly there is no one right answer. This is due to a number of factors including these: the bills can be stacked different ways in the box; they can be folded in the box; they can be pressed tightly down or left loose. What is important is that students measured and calculated accurately and found an approach that was logical, sensible, and yielded a reasonable answer.

When all solutions have been presented, it is finally time to open Pete's box. As you open the box, be sure to find Pete's poem first and read it aloud to the class with surprise. Distribute the candy reward to each student. Remind students that nobody said there was money in the box. The ranger was merely speculating. Apparently he did not know Two-Dollar Pete as well as he thought!

CUSTOMIZING THE PROJECT

The problem was originally designed for New Jersey students. You may wish to replace the first paragraph and have Ralph find the box in a familiar local state or national park, forest, or wilderness region.

Some groups will finish this problem rather quickly. In fact, some second graders may successfully complete the project by Day 3. You should probably plan for only three or four days with third and fourth graders.

To simplify this problem, you can alter the size of the box. A smaller box that will fit bills exactly will make the problem easier. Letting individuals or partners work on the problem (instead of groups of three to four) will allow some students more time to think through and develop a strategy that works for them and not just for their group.

ASSESSMENT

As you review student work and the process you observed, consider these questions:

Is measurement accurate?

Is calculation accurate?

Is the solution a reasonable estimate by your criteria?

Was the data well-organized, labeled, and clear?

Are the strategies logical, reasonable, and efficient?

What kinds of mathematics were used?

Was a calculator used? If so, how?

For additional tips on assessment see Appendix 1, Assessment, pages 97-99.

Math Journal Writing or Interviews

Ask students to answer one or more of the following questions either in writing or orally:

What mathematics did you use to solve Park Ranger Ralph's problem?

Why was it important in this problem that Pete stole only two-dollar bills? What did you do about this?

Why do you think different groups got different answers, even though they used the same methods?

What strategy worked particularly well for you in solving this problem?

CURRICULUM CONNECTIONS

Literature Larger numbers are hard to conceptualize. This problem took students into the thousands, but *How Much is a Million?* by David M. Schwartz does a good job of helping students conceptualize what a million is really like.

Math Ask students how they would find out how many bills would fit in a different box. Would they have to start all over, or is there information they now know that could help them solve new problems like this one?

Art Have students draw a wanted poster of Two-Dollar Pete.

Music Lots of songs have been written about money. Ask students to find or name songs that talk about money.

Language Arts Have students write *The Legend of Two-Dollar Pete* that includes the following information: What is he like? Where is he from? Why does he have the strange habit of stealing only two-dollar bills?

Social Studies The pine barrens of southern New Jersey is a real and interesting place. There are many legends and a great deal of history. Read selections from *Blackbeard the Pirate and Other Stories of the Pine Barrens* by Larona C. Homer. If you have used a different locale for the project, look for information for students about the history or folklore of the locale.

Science Ask students whether they would ever need to find out volume or capacity in a science lab. Ask which tools or ideas in math are helpful for finding volume and capacity in science.

Dear Students,

While patrolling recently in a cranberry bog deep in the New Jersey Pine Barrens, I came upon this mysterious, sealed box, sunk in the bog.

I have measured the box with great accuracy. It is 12 inches wide, 16 inches long, and 2 inches high.

I believe there may be money inside. In fact, I am almost positive it belonged to the notorious Two-Dollar Pete, a bank robber who had the strange habit of stealing only two-dollar bills.

I would like you to open the box and find out what is inside. But first, because I love math as much as you do, I'd like to challenge you to figure out the answers to these questions:

1. How much money do you think Pete could fit into the box?
2. If you shared that amount of money with the members of your class, how much would each person get?

No matter what your answers are, be sure to show clearly how you got them.

      Good Luck,
      Park Ranger Ralph

P.S. Once you figure out the problem, you may open the box. Whatever is inside is yours to keep with my compliments.

So you found my box
And you figured it out,
Well that makes me feel just dandy.
But if it's my loot you're after,
Sorry to say,
This time you only get candy!

PETE

TWO-DOLLAR PETE

Dear Students,

We are the owners of the Haunted Mansion Fun House Company. We have bought land near your school, and we would like to build a new Haunted Mansion Fun House. Right now, we are in the planning stage and could use your help. We want this mansion to be the spookiest and most fun possible. Would you design a layout for the exhibits in the mansion and make a map showing your design?

Here is the information you will need.

1. The new mansion will be 24 yards long, 15 yards wide, and one story high.

2. All the exhibits are rectangular and are 5 yards wide, but their lengths vary.

3. All the exhibits must fit in three rows. Each row will be 5 yards wide.

4. The mansion will have one entrance and one exit. On your map, use arrows to show the way people will enter the mansion, walk through all the exhibits, and leave the mansion.

5. Each exhibit takes a certain amount of time to view. We want our guests to spend at least 20 minutes in the mansion.

6. Use a scale of 1 inch = 1 yard when drawing your map.

7. The moving van company that will bring the exhibits to the new mansion has told us that the total weight of the exhibits must be less than 600 pounds.

Attached is a list of exhibit data that includes lengths, weights, and viewing times. If there is room on your map, create some exhibits of your own. Just remember that customers must spend at least 20 minutes in the mansion, and that the total weight of the exhibits must be less than 600 pounds.

Thank you in advance for your kind help! We look forward to seeing your designs.

Sincerely,

Rather & Veri Strange

Rather and Veri Strange

HAUNTED MANSION

Grade Level: 3

Overview

Mr. and Mrs. Strange, owners of the Haunted Mansion Fun House Company, send the class a letter asking for its help in designing a new Haunted Mansion. In response, students create floor plans showing the location of the exhibits described in the letter. In meeting the size, time, and weight requirements set forth by the Stranges, children invent additional exhibits.

This problem can work well with second graders. See Customizing the Project, page 80.

Objectives

- To make a map or plan according to a given set of data
- To control three variables (time, measurement, weight) while creating the map or plan

The Mathematics

- *addition*
- *subtraction*
- *multiplication*
- *measurement*
- *time*
- *weight*
- *unit conversion*
- *scale*
- *scale drawing*
- *map-reading*
- *spatial reasoning*
- *controlling multiple variables*

Time Needed   four to five 40-minute periods

Grouping   groups of three to five students

Preparation and Materials

Day 1

- Preparation and Reference Sheet (optional), 1 per student, page 84
- Haunted Mansion letter and Exhibit Data sheet, 1 per student, pages 82-83
- Large envelope to hold letters and data sheets (Decorate with spooky drawings.)

Days 2, 3, and 4

- Haunted Mansion time and shipping weight lists (optional), 1 per group, pages 85-86
- Large sheets of drawing paper, several sheets per group
- Tape
- Rulers, at least 1 per group
- Calculators (optional), at least 1 per group
- Class dictionary, 1 per group
- Social Studies books (or any other class text), 1 per group
- Scales (bathroom and postal), 1 of each for groups to share

Day 5

- Thank-you cards (optional), 1 per student, page 87

PROJECT: HAUNTED MANSION

Day 1: Special Delivery

You may want to have an accomplice deliver the letters during class.

Deliver

Make a good show of "discovering" a mysterious-looking envelope that is covered with spooky drawings and addressed to the class. One way to do this would be to place the envelope some place in the classroom where a student is likely to notice it. Open the envelope for the class and pass out the letters and exhibit data sheets.

Read

Give students time to read the letter silently. Then have them take turns reading aloud. Discuss any vocabulary and terminology that might be new to students.

Discuss

The problem is purposely woven into the context of the letter. To help students interpret the letter and their task, ask

1. So, what's the problem?

Use some or all of the following questions to focus the discussion and analysis of the letter.

Who sent you the letter? (Mr. and Mrs. Strange, owners of the Haunted Mansion Fun House Company)

What have Mr. and Mrs. Strange asked you to do? (To make a map showing a design for a new Haunted Mansion Fun House)

Have you ever been in a Haunted House at an amusement park? What was it like? (Allow students to share what they know to build a connection to previous knowledge and to make the problem more motivating.)

Are there guidelines? (Yes)

What are the guidelines? (List volunteers' responses on the chalkboard.)

When students have finished responding to the last question, you may need to ask these additional questions to be sure that students have identified all the pertinent information.

What do the guidelines say about

You may want to go through the list of exhibits, one by one, and review the information given about each.

- *the size of the mansion?* (It's 24 yards long, 15 yards wide, and 1 story high.)
- *the entrances and exits?* (There is one of each. We need to use arrows to show how people will walk through the exhibits.)
- *the amount of time customers will spend in the mansion?* (At least 20 minutes)
- *the arrangement of the exhibits?* (They are in three rows; each row is 5 yards wide.)
- *the number of exhibits?* (The list has ten. We may add some.)
- *the width and length of the exhibits?* (Each is 5 yards wide; the lengths vary.)
- *the weight of the exhibits?* (The total must be less than 600 pounds. The moving van company needs to know the total weight.)

- *the scale of the maps you will draw?* (1 inch = 1 yard)
- *creating exhibits of your own?* (We can do this if we have arranged all ten exhibits, there is still room, and we meet the time and weight conditions.)

Once you have reviewed all the information students need, ask

2. What materials do you think you will need to solve this problem?

List volunteers' responses. Tell students you will gather the necessary materials they have requested so that they can start work the next day.

This could be a good time for a mini-lesson on scale or a discussion of a sample map or floor plan with the scale clearly indicated.

Day 2: Let's Get to Work!

Have all materials ready for students and form the working groups. Decide whether you want students to use the optional time and weight lists. Briefly review the questions and information from Day 1. Ask

> *Are there any questions or concerns about what you are expected to do?*

Ask students to record the total time and weight somewhere on their finished map and to save all relevant work and calculations. Once you are satisfied that students understand the task and the parameters, have them begin work.

Using the time and weight lists precludes students from creating their own organizational strategies.

Common Student Approaches

Many groups will begin by creating the correct size paper. This might involve cutting and/or taping paper together, depending on the size of paper you supply. Other groups may begin by totaling data from the exhibits list or by trying to make subtotals of 24 yards from the list. If a group appears to be stuck, you may wish to revisit the guidelines with that group and offer suggestions for getting started (such as the approaches mentioned above) or other hints that will get them started without "giving away" a complete strategy.

Teacher Role

While students are working, circulate, assist, and observe. In addition, use this time to begin assessment. Keep these assessment questions in mind:

> *What math are students doing?*
>
> *What organizational ideas are the groups using?*
>
> *How is the group cooperating and dividing up tasks?*
>
> *What are students doing well?*
>
> *What difficulties are students having?*
>
> *What tools are students using?*

Day 3: Time to Share

Begin with a whole class discussion to give students the opportunity to evaluate their partial solutions and consider the ideas and suggestions of others. Have students sit in a circle with their work. Ask the following questions and discuss answers together:

> *What did your group do first? Why? Was it a good choice?*
>
> *What has worked out well for you?*

What problems have you had? Do other groups have suggestions regarding these problems?

How have you worked out the sizes of the exhibits on your map?

How are you keeping track of what you are doing? How do you know what you have done so far?

If you have put in a few exhibits on your map, do you know how much time you have used? How are you keeping track?

How do you make exhibits fit in the rows on your map?

After the discussion, direct students to return to their groups and continue working. Again, as you did yesterday, circulate, assist, observe, and assess.

Day 4: Still Working

Begin with a whole class discussion. Have groups display their partially finished maps. Ask the following questions:

What has worked well for you?

What problems have you had? How are you trying to solve these problems?

Before groups continue their independent work, remind them to be ready to present the finished maps for the next class. While students work, circulate, observe, assess, and assist.

Day 5: Closure

Have groups display their maps and take turns explaining their designs to the class. Use these questions to help summarize the week's work:

Was there any one right way to make the map? Explain.

How can you tell if a map meets the guidelines in the Stranges' letter?

If you wish, have Mr. and Mrs. Strange "send" students the thank-you card.

CUSTOMIZING THE PROJECT

This is a good project to do around Halloween when everyone is "in the mood," but it can really be done anytime. You might like to use this as a way to explore measurement or as a culminating problem after a unit on measurement.

You may want to work through the optional Preparation and Reference Sheet before beginning. This can be a handy reference for students as they do the problem.

There is lots of room to make this problem "just right" for your particular class while keeping the basic task the same. The key to customizing is in the details of the time, dimensions, and weight of the exhibits. You can make the problem easier or harder or make it reflect the units of measure your class is studying. For instance, make all the units metric and eliminate the conversion of customary units. Give out the shipping weight and time forms to help students organize their work. Or allow students to come up with their own way of organizing the data.

ASSESSMENT

As you evaluate students' work, consider these questions:

Does the map include all the given exhibits?

Is the map clear and readable?

Is the scale of the map accurate? Does it match the data provided?

Is the total exhibit time at least 20 minutes?

Is the shipping weight total accurate? Is it less than 600 pounds?

Is the data well organized?

For additional tips on assessment see Appendix 1, Assessment, pages 97-99.

Math Journal Writing or Interviews

Ask students to answer one or more of the following questions either in writing or orally:

What mathematics did you use to help Mr. and Mrs. Strange?

How did you keep track of the time, size, and weight of the exhibits?

What is scale? Give an example.

What strategy worked particularly well for you in solving this problem?

What difficulties did you find, and how did you get around them?

CURRICULUM CONNECTIONS

Literature *In a Haunted House*, by Eve Bunting, is an easy read and has a cute twist of an ending. The exhibits in this house are varied and interesting. If the reader looks closely, he or she can see in the illustrations just how the "spooky exhibits" have been created.

Math Have students create admission price guidelines for the Haunted Mansion. Remind them to consider the age of customers, the time spent in the mansion, and how often a customer might visit.

Art Have students make a three-dimensional model of one of the Haunted Mansion exhibits they created using a diorama, sculpture, or pop-up scene format.

Music Listen as a class to some spooky music, such as *Danse Macabre* by Saint-Saëns or the *Toccata in d Minor* by Bach. Discuss what makes the music sound spooky.

Language Arts Have students write an advertising brochure for the Haunted Mansion that includes the names of some of the attractions and describes the fun and excitement people could have when visiting the mansion.

Social Studies Have students bring in maps from home and have a "Map Show and Tell" in which students find the key (including the scale) on each map and then discuss the meaning of a key and what it shows.

Science Have students write a spooky message with "invisible ink." See *A Teacher's Science Companion* by Dr. Phyllis Perry for details on this experiment.

Dear Students,

We are the owners of the Haunted Mansion Fun House Company. We have bought land near your school, and we would like to build a new Haunted Mansion Fun House. Right now, we are in the planning stage and could use your help. We want this mansion to be the spookiest and most fun possible. Would you design a layout for the exhibits in the mansion and make a map showing your design?

Here is the information you will need.

1. The new mansion will be 24 yards long, 15 yards wide, and one story high.
2. All the exhibits are rectangular and are 5 yards wide, but their lengths vary.
3. All the exhibits must fit in three rows. Each row will be 5 yards wide.
4. The mansion will have one entrance and one exit. On your map, use arrows to show the way people will enter the mansion, walk through all the exhibits, and leave the mansion.
5. Each exhibit takes a certain amount of time to view. We want our guests to spend at least 20 minutes in the mansion.
6. Use a scale of 1 inch = 1 yard when drawing your map.
7. The moving van company that will bring the exhibits to the new mansion has told us that the total weight of the exhibits must be less than 600 pounds.

Attached is a list of exhibit data that includes lengths, weights, and viewing times. If there is room on your map, create some exhibits of your own. Just remember that customers must spend at least 20 minutes in the mansion, and that the total weight of the exhibits must be less than 600 pounds.

Thank you in advance for your kind help! We look forward to seeing your designs.

Sincerely,

Rather & Veri Strange

Rather and Veri Strange

EXHIBIT DATA

These exhibits must be in the Haunted Mansion. The time shown after each exhibit tells how long it will take a visitor to walk through the exhibit and move on to the next one. Remember, each exhibit is 5 yards wide.

1. Meet the Wolfman (2 minutes)
   length: 15 yards; weight: 100 pounds

2. Ghosts Flying (3 minutes)
   length: 5 yards; weight: 35 pounds

3. Swamp Thing Rising (1 minute)
   length: as long as the teacher's desk; weight: 56 pounds

4. One-Eyed Monster (2 minutes)
   length: 4 yards; weight: 16 ounces

5. Hall of Crazy Mirrors (1 minute)
   length: as long as 2 chalkboards; weight: ten times as much as a class dictionary

6. Skeletons (30 seconds)
   length: 3 feet; weight: 1 pound

7. Bats Flying (1 minute)
   length: 10 yards; weight: 32 ounces

8. Creaky Door (30 seconds)
   length: as long as the width of a classroom door; weight: as much as the oldest student in the class

9. Chains Rattling (2 minutes)
   length: as long as the width of four student desks; weight: as much as 50 Social Studies books

10. I Want My Mummy! (1 minute)
    length: 5 feet; weight: 10 kilograms

ADDITIONAL EXHIBITS AND DATA:

_____

_____

_____

_____

_____

_____

Preparation and Reference Sheet

Abbreviations

inches = _____

feet = _____

yards = _____

meters = _____

centimeters = _____

ounces = _____

pounds = _____

grams = _____

kilograms = _____

Conversions

_____ in. = 1 ft

_____ ft = 1 yd

_____ cm = 1 m

_____ oz = 1 lb

_____ g = 1 kg

_____ lb = 1 kg

A dictionary weighs about _____.

A social studies book weighs about _____.

How long is a ruler? _____

How would you weigh a person? _____

What units would you use to measure the width of a student desk?

_____

What units would you use to measure the length of a chalkboard?

_____

How many minutes are in an hour? _____

How many seconds are in a minute? _____

Haunted Mansion Time List

Exhibit	Time
_____	_____
_____	_____
_____	_____
_____	_____
_____	_____
_____	_____
_____	_____
_____	_____
_____	_____
_____	_____
_____	_____
_____	_____
_____	_____

Total Time _____

Haunted Mansion Shipping Weight List

Exhibit

_____

_____

_____

_____

_____

_____

_____

_____

_____

_____

_____

_____

_____

_____

Weight

_____

_____

_____

_____

_____

_____

_____

_____

_____

_____

_____

_____

_____

_____

Total Weight _____

You can print this in greeting card format with a piece of reflective foil taped in the middle so that students see their reflections when opening the card.

THANK YOU

Because we care enough to send the strangest.

Dear Students,

I saw a mummy
I saw a ghost,
I saw lots of spooky things,
But YOU scared me the most!

Tape foil here

Thanks again for your help,
The Stranges
The Stranges

Fold

GREETINGS EARTHLINGS!

We come in peace, seeking your assistance. We are the inhabitants of Planet Zignox. We are quite small compared to you. Our tallest Zignoxian is 1 cm high. (He also happens to be our best basketball player.)

We are about to build our planet's most important edifice, our first school. We are searching the universe for architectural ideas. Would you help us by building a model of a school for us? As you can guess, it will be a small school compared to yours.

Please use Cuisenaire Rods, Base Ten Blocks, or a combination of the rods and blocks for your models. You can hold the rods together with that amazing earth invention - tape! Please tape your model onto a cardboard base.

The base of the school must have a perimeter that is no greater than 80 cm. The building should not be higher than 100 cm at any point. The building should have at least one entrance and one exit. We also like windows.

There is one more thing you need to know. We have a budget to work with. We only have so much to spend you know! Each centimeter of building material will cost us 1 ziggie. (A ziggie is our unit of money, something like your U.S. dollar.) For example,

> 1 white rod costs 1 ziggie to us.
>
> 1 red rod costs 2 ziggies.
>
> 1 orange rod costs 10 ziggies.

If you are using Base Ten Blocks, the unit block costs 1 ziggie, the tens block costs 10 ziggies, and the hundreds block costs 100 ziggies.

The total cost of your building has to be between 1,000 and 2,000 ziggies.

We are counting on you to do the best job you can. We will transport ourselves down to inspect your work in about five days. You will be richly rewarded if you follow our rules and make an interesting model!

Good luck and **"Jobulizzy Zirkinoxee!"**
(English translation: Have Fun!)

Sincerely,

**YOUR FRIENDS
FROM PLANET ZIGNOX**

PLANET ZIGNOX

Grade Level: 3

Overview

A mysterious saucer-shaped object is found on the playground. Inside is a request for help from the friendly aliens from Planet Zignox where the tallest inhabitant is only 1 cm tall. Zignoxians are small, but they have big plans! They are designing their planet's most important edifice - a school. Where better to get ideas and help than from students in a school on Earth? Students help the Zignoxians by building models that conform to the conditions set forth in the request.

This problem can be successfully completed by second graders and yet be made challenging for fourth graders. See Customizing the Project, page 93.

Objectives

- To control two variables (money and measurement) simultaneously while solving a complex, real-life problem
- To apply a variety of math concepts within given parameters to create a physical model

The Mathematics

- *perimeter*
- *repeated addition*
- *double-digit addition*
- *place value*
- *multiplication*
- *linear measurement*
- *calculator skills*
- *money*

Time Needed four to five 40-minute periods

Grouping This is an ideal project for partners, but groups could have as many as three or four students.

Group size will most likely be determined by the number of class sets of Cuisenaire Rods you can obtain. The more the better!

Preparation and Materials

Day 1

- Letter from Zignoxians, 1 per student, page 95
- A "flying saucer" to hold the letters (Use two pie plates. Place the letters in one plate and place the other plate upside down on top of the first to create a saucer shape. Cover the shape with foil and seal with tape. Use a marker to draw a large "Z" on the foil and decorate the rest of the saucer to indicate its extraterrestrial origins. This makes a good, basic saucer. No doubt you can come up with a more imaginative space ship for delivering the problem!)

Days 2, 3, and 4

- Cuisenaire® Rods, 1 class set per 12 students
- Base Ten Blocks (optional), 1 class set per 24 students
- Centimeter rulers, 1 per group
- Calculators (optional), 1 per group
- Cardboard, one 12 in. × 18 in. piece per group

Building with Base Ten Blocks is quicker and easier and allows students to erect floors and walls rapidly.

- Transparent tape, 1 roll per group
- Scissors, 1 pair per group (Small pieces of transparent tape seem to work best for holding models together, and students like to use scissors to cut the tape into very small pieces.)

Day 5

- A second flying saucer
- Contents of the second flying saucer
  - ◆ Lots of ziggies, the money used on Planet Zignox (These can be very small pieces of cut paper with Z's prominently printed on them.)
  - ◆ An audio cassette with a message from the Zignoxians (Record a message in a Zignoxian voice. Apologize to students that you and your crew of Zignoxians came back while they were out, but that you are pleased with the models and are leaving the ziggies as a reward.)
- An audio cassette player

PROJECT: PLANET ZIGNOX

Day 1: Special Delivery

Deliver

"Discover" the mysterious object, the flying saucer, in the schoolyard and bring it into the classroom. It might be fun to have an accomplice discover the saucer outside and come up to your window holding it. Or, you may want to return from lunch or gym and find it lying in the middle of your classroom floor. Open the package slowly and carefully. Remember, it's a strange object that could be from another planet. Ask students what they think it might be. Distribute one letter to each student.

Read

Give students time to read the letter silently. Then have them take turns reading aloud. Discuss any vocabulary and terminology that might be new to students.

Discuss

All of the information needed to solve the problem is in the letter. During discussion, it is important that students get a good understanding of just what the problem is and what they are required to do. Ask

1. So, what's the problem?

Use some or all of the following questions to focus the discussion and help students understand the letter and the problem.

Who sent you the letter? (Zignoxians from Planet Zignox)

What do you know about other planets? (Allow some time for students to discuss and share what they know about planets in our solar system and other knowledge about space. This helps build a connection to previous knowledge and makes the problem more motivating and fun.)

What do the Zignoxians want us to do? (Build a model of a school for them.)

What do we use to make the model? (Cuisenaire Rods and/or Base Ten blocks built onto a cardboard base and held together with tape)

Do the rods have a value? (Yes, each centimeter of rod has a value of 1 ziggie. So one 1 white rod = 1 ziggie, 1 red rod = 2 ziggies, and so on.)

How tall is the tallest Zignoxian? (1 cm)

How big should the model be? (It cannot be higher than 100 cm and must have a perimeter that is no greater than 80 cm.)

What is perimeter? (The distance around the edges of an object or shape)

A mini-lesson about perimeter may be appropriate here. The need to know will motivate some very focused attention as you teach the mini-lesson.

Are there any other limits or requirements the model must have? (It must have a separate entrance and exit. Windows are suggested but not required. The model must have a total value that is between 1,000 and 2,000 ziggies.)

What will happen when the model is done? (The Zignoxians will transport themselves down and richly reward the students.)

To help students gain a sharper focus on what they are to supposed to do, ask

What makes a good school building?

This question gets students to connect the problem to their own experience. Discuss the physical features of schools such as windows, doors, hallways, and shape. Talk about your own school and the features that students like and dislike. List students' ideas on the board. Urge them to think about these ideas as they build the school for the Zignoxians.

When you feel that students have all the information they need, ask,

2. What materials do you think you will need to solve this problem?

List responses on the board. Ask students to justify their thinking regarding the need of each material. This process helps students think about strategies and approaches.

Do you want students to use calculators? If you want students to practice and apply the concept of double-digit addition with regrouping, you may *not* want them to use calculators. If this concept is not a primary objective, this problem may be a good opportunity for showing students that a calculator is an appropriate tool for adding large numbers, keeping track of totals, and/or checking addition.

Assure students that necessary materials will be made available to them when they begin.

Day 2: Let's Get to Work!

Have all materials ready for students. Form partner or small groups. Briefly review the questions from Day 1 so everyone is clear on task and parameters. Ask

Are there any questions or concerns about the problem?

Do you all know what you are required to do?

Once you are satisfied that students have a reasonable understanding of the task, have them begin work.

Common Student Approaches

Many students begin by "measuring out" a perimeter. They then build in layers up from the perimeter. For some, the color of the rods is important, and they will use the rods in a patterned manner to enhance the look of the model. For others, color is a non-factor. Once groups decide on a perimeter, they often work on the cardboard base, sometimes cutting it to the exact perimeter.

A common pitfall for students is that they begin building and do not know how much they are spending. As they build, they are spending money, just like in real life. They need to develop a system as they proceed that will allow them to keep this data organized and let them know at any given point how much money they have spent. Remind students that they really need to think about "keeping track" of numbers in this problem.

Teacher Role

While students are working, circulate, assist, and observe. Use this time to begin assessment. Keep these kinds of assessment questions in mind:

What math are students using?

How is the group cooperating and dividing up the tasks?

What strategies are students using?

How are these strategies developed?

How efficient are the strategies?

What are students doing well?

What difficulties are students having?

What tools are students using?

Day 3: Time to Share

Begin with a whole class share session. Have partners sit in a circle with models displayed in front of them. Encourage students to listen for good ideas during discussion. Ask the following questions and discuss answers together:

What did your group do first? Why? Was it a good choice?

What has worked well for you?

What problems have you had?

What have you done about the perimeter?

Have you kept track of money spent? How?

As answers are given, take advantage of the opportunity to reinforce valuable concepts or ideas students suggest. After discussion and shared answers, have students begin work. Again, circulate, assist, observe, and assess.

Day 4: Still Working

Begin with a whole class share session and a shared display of "models in progress." Ask the following questions and discuss:

What has worked well for you?

What problems have you had?

How are you trying to solve these problems?

After discussion and shared answers, have students return to work. As before, circulate, observe, assess, and assist.

Day 5: Closure

Have each group display their completed models and tell about their work: what sort of process they went through; what physical features of the model the class should be aware of; how they kept track of money and measurement; what materials they chose and why; how they divided up the workload.

Place all the models on display. Arrange for another pie-plate "flying saucer" to arrive while students are out of the classroom. When students return to class, open it and play the audio cassette. Then distribute the ziggies. These little pieces of paper should be redeemable with you for books, erasers, pencils, prizes, or whatever is an appropriate reward for your particular class.

CUSTOMIZING THE PROJECT

There is a good deal of flexibility in the parameters of the problem. You can adjust the budget and change the total number and range of ziggies. Tightening the range will make the problem more challenging. You can increase or decrease the perimeter requirement or eliminate the idea of perimeter altogether. To make the problem quicker and easier, you can have students use only Base Ten Blocks which will give them building sections which are worth 100 ziggies apiece.

ASSESSMENT

Examine each completed model in terms of the criteria of the letter:

> *Is the perimeter 80 cm or less?*
>
> *When the values of the rods are added together, do they have a total value that falls between 1,000 and 2,000 ziggies?*
>
> *Is there an entrance and an exit?*
>
> *Are there any windows?*

For additional tips on assessment, see Appendix 1 Assessment, pages 97-99.

Think about the process you observed over the course of the project and consider these questions:

> *What kinds of mathematics did students use?*
>
> *Was measurement accurate?*
>
> *Was calculation accurate?*
>
> *Was data well-organized?*
>
> *What kinds of methods and strategies were used?*
>
> *Were methods logical, reasonable, and efficient?*
>
> *Was a calculator used? If so, how?*

Math Journal Writing or Interviews

Ask students to answer one or more of the following questions either in writing or orally:

What mathematics did you use to solve the problem and complete the model?

What is perimeter?

How did you keep track of the amount of money you spent?

If you were to share one strategy or method that worked particularly well for you in solving this problem, what would it be?

CURRICULUM CONNECTIONS

Literature *The Magic School Bus; Lost in the Solar System*, by Joanna Cole, is about a wild class trip through our solar system taken by Miss Frizzle and her students.

Math With your students, measure the actual perimeter of your school or class. For homework, have students measure the perimeter of their house or bedroom.

Art Invite students to create a Zignoxian flag, banner, or logo. Ask them to draw what a Zignoxian might look like.

Music Ask students to write a Zignoxian National Anthem or a school song for their new school. Gustav Holst's musical composition, *The Planets*, has a theme for each planet in our solar system. Have students choose and research a planet and then listen to its theme, and discuss how the theme is appropriate for that planet.

Language Arts Have students write about life on Zignox, or ask them to name the new school, and explain why the name was chosen.

Social Studies Have students investigate how real schools in their community are built and paid for.

Science Remind students that there is no real Planet Zignox. Ask them to choose one of the real planets in our solar system and find out what it is actually like.

GREETINGS EARTHLINGS!

We come in peace, seeking your assistance. We are the inhabitants of Planet Zignox. We are quite small compared to you. Our tallest Zignoxian is 1 cm high. (He also happens to be our best basketball player.)

We are about to build our planet's most important edifice, our first school. We are searching the universe for architectural ideas. Would you help us by building a model of a school for us? As you can guess, it will be a small school compared to yours.

Please use Cuisenaire Rods, Base Ten Blocks, or a combination of the rods and blocks for your models. You can hold the rods together with that amazing earth invention - tape! Please tape your model onto a cardboard base.

The base of the school must have a perimeter that is no greater than 80 cm. The building should not be higher than 100 cm at any point. The building should have at least one entrance and one exit. We also like windows.

There is one more thing you need to know. We have a budget to work with. We only have so much to spend you know! Each centimeter of building material will cost us 1 ziggie. (A ziggie is our unit of money, something like your U.S. dollar.) For example,

   1 white rod costs 1 ziggie to us.

   1 red rod costs 2 ziggies.

   1 orange rod costs 10 ziggies.

If you are using Base Ten Blocks, the unit block costs 1 ziggie, the tens block costs 10 ziggies, and the hundreds block costs 100 ziggies.

The total cost of your building has to be between 1,000 and 2,000 ziggies.

We are counting on you to do the best job you can. We will transport ourselves down to inspect your work in about five days. You will be richly rewarded if you follow our rules and make an interesting model!

Good luck and **"Jobulizzy Zirkinoxee!"**
(English translation: Have Fun!)

Sincerely,

**YOUR FRIENDS FROM
PLANET ZIGNOX**

Jobulizzy Zirkinoxee!

FROM HAUNTED MANSION
PAGES 76-87

Appendix 1: Assessment

There is not room here, nor do I have the expertise or inclination, to thoroughly discuss current assessment practices. Instead, here is a brief overview of assessment specifics related directly to the types of problems in this book. Project Problems lend themselves well to many forms of assessment, and they can enable you to gain a great deal of information about students and their learning.

There is no one "right" form of assessment or tool that does it all. Different tools give different kinds of information. Consider using the three methods discussed here in conjunction with one another to get as complete a picture as possible of students' learning.

Observation Checklists

When evaluating the process of Project Problem work, good observation is paramount. Fortunately, the nature and structure of Project Problems lend themselves to freeing up the teacher to move easily about the room to observe, question, and guide. While in this role, the teacher can use observation checklists. For those who like a checklist, consider including the following observation cues.

Is the student participating? How?

What is the student actually doing?

What is the student doing well?

What is the student having problems with?

Has the student assumed a role in the group? What is it?

What sorts of strategies are being used? Are they efficient? sensible? accurate?

What kind of math is being done? Describe.

How accurate is computation?

What does the student do when encountering difficulty?

How would you judge the student's effort?

If there is a product,

What were the expectations for the product?

Does the product meet the client's needs and specifications?

Does the product meet all the parameters outlined in the problem?

Rubrics

Rubrics can help organize thinking about assessment and guide observation and recording. There are many generic rubrics available from a wide variety of sources. Here is a sample I developed to use with Project Problems in my classroom. There are six areas discussed in this rubric: problem solving strategies, mathematical accuracy, completeness, cooperation, presentation, and independence. The highest score is 5, and the lowest is 1. I never use this rubric as an assessment tool on its own. Rather, I use it in conjunction with other pieces such as interviews, journals, and observations checklists, so that other areas, like the kinds of mathematics used, can be considered.

Circle the phrase in each area that best applies to the work being assessed.

5 Creative, efficient strategies used

Math is completely accurate

All work is shown

Very cooperative; listens and works together with others

Very neat, very clear, well written answers and presentation

Independent

4 Very effective use of strategies

Math is almost completely accurate

Most work is shown

Cooperative; works well with others most of the time

Answers are neat, clear, and understandable

Minimal teacher help required

3 Some effective use of strategies

Math is mostly accurate

Some work is shown

Sometimes cooperates with others

Answers are readable but not completely clear

Teacher help on some parts

2 Strategies tried but not effectively

Math is mostly inaccurate

Very little work is shown

Does not cooperate with others

Answers are not readable, not at all clear

Teacher help on all parts

1 No strategies used

Math is completely inaccurate

No work is shown

Does not participate or cooperate with others

No answers given

Refused to work alone or with teacher

Math Journal and Interviews

A very good way to get insight into how and what students are thinking about math ideas and concepts is to have them write about their thinking. It is not an easy task for students to write about math, but it is a very good way to get them to think clearly and reflect carefully on what they have actually done and learned in their problem solving. Those who cannot yet write can talk about it.

Each of the problems in this book includes problem-specific questions for math journal responses. The individual discussion of math journal responses and artifacts from the problem solving can form an excellent basis for student interviews. In these interviews students have an opportunity to explain and elaborate on the thinking shown in their written work. This is a form of assessment that can provide invaluable insight into the thought processes and understanding students have about math.

A Journal Supply Tip: A few years ago I was looking through one of the larger school supply catalogs. I noticed that the college exam blue books we all remember and love were very cheap – cheaper than regular spiral-bound journal books and cheap enough so that I could buy each student a book for every Project Problem. I was amazed at how many I could get for just a few dollars, so I bought them and have been reordering every year since.

Appendix 2: Project Problem Development

I hope that if you've tried the Project Problems in this book, you've found them successful and fun. So much so, that you'll want to do more each year. My experience is that after completing two problems, my students begin to live from problem to problem, asking me constantly when the next one is coming. So, if you've had a good experience, you may want more. Now, where to get them?

The very best source is not going to be my next book (though, of course, I wouldn't discourage you from looking there) or anyone else's, but rather it will be you and your colleagues. Teachers can and do develop these problems. Problems that you develop yourself or with colleagues will be more specific, engaging, and fitting to your style and curriculum than any you will find elsewhere. Is it an easy task? No, BUT there are some ideas that can help.

The technique I use with groups of teachers for problem development revolves around something I call, for lack of a better term, the Development Triangle. As with all the triangles I've ever met, the Development Triangle has three legs.

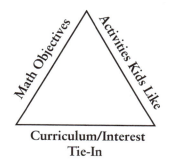

1. Math Objectives
2. Activity Kids Like
3. Curriculum or Interest Tie-In

Looking at these three facets of a problem can give you a start. First, think of a math objective or concept you want to address. The complex nature of these problems or most any real life problem rarely leaves you with one solitary objective. There are many concepts involved in a given problem, but I try to make sure there is an overriding concept that I can observe my students apply and engage in. Look to your math curriculum guide. What are the major concepts at your grade level? I'm presently teaching second grade. Major concepts that I've written problems on so far are: place value, time, linear measurement, money, fractions, addition, and use of calculators. Begin by identifying one major concept or math theme you want to address. Also, start to think about how this math concept is used in real life. In what ways? By whom? Will your students need to use it in their lives? How?

Second, consider what students at your grade level like to do. What sorts of activities are engaging to your kids? I've made lists that I work from. My K-3 list includes things like painting, drawing, clay work, weaving, model building, miniatures, stickers, board games, stuffed animals, acting, and singing.

Third, think about ties to other parts of your curriculum or student interests. This might form the backdrop or setting of your problem. When developing *Planet Zignox*, the idea was tied to Space and Planets. This was a part of our science curriculum, but it was also an area full of natural interest for kids at this age. Other third Developmental Triangle "legs" I have used at these levels include holidays (like Halloween), sports, fairs, carnivals, rides, money, rain forests, and zoos.

Play around with the second and third legs of the triangle. Does a scenario suggest itself? How would your math concept be used in this situation? Could it be used here?

There are some people who can do this sort of problem creation on their own. Most of the rest of us work better in a team. Get together with other interested teachers at your grade level. (Use the same poor individuals you coerced to work through the problems back in Chapter Three.) Begin by discussing and drawing up your three lists, one for each leg of the triangle. Take your time. Bring food, sit, and talk.

Now that you have your lists, try to fill in the three legs of the triangle. Juggle the ideas on the three legs. Write down everything everybody suggests. Throw no idea away. One of these bits may later form the basis of your best problem. Just brainstorm. Don't critique yet. If you work at it long enough, a kernel of an idea that works will emerge. Write it down and flesh it out. How will you introduce it? Who is the client? What do students have to do? Is it fun? What is the task? How will you assess it? Write it up and give it a try.

Another approach that has been successful for problem developers is looking at newspapers and magazines for stories and articles that involve math. This in itself is an interesting exercise for students and a great homework assignment. Articles from newspapers might provide a source of a math objective, a scenario, and a curriculum tie all in one place. Then it's just a matter of deciding what you will ask the class to do. What is the problem you can draw from the article?

A good example of this sort of problem creation is a story that was in a New York newspaper about a bank robbery. Bank officials stated that a robber had carried off a million dollars in small bills. Some readers with an eye on the math, in particular the weight and volume of a million dollars in small bills, wondered if it was indeed possible for a person to actually do this. Educational Testing Service developed a Middle School assessment problem from this story and called it, *Million Dollar Getaway*.[1] A problem in this book, *Two-Dollar Pete*, is a derivative of this same idea.

Does this seem daunting? Yes?! Admittedly it's not easy, but it can be done, and the reward is great. The more teachers you can get to join your development group the better. Talk freely, discard no idea. You'll find the scenario that fits your kids and teaches what you want to teach. Your students will be learning math in a real, fun, engaging, and meaningful way, and you and your Project Problems will be a large part of the reason why!

[1] *Packets Performance Assessment For Middle School Mathematics; Million Dollar Getaway*, Educational Testing Service, Princeton, NJ, 1995.

Appendix 3: References

Curriculum and Evaluation Standards for School Mathematics, National Council of Teachers of Mathematics, Reston, VA, 1989.

Equity, Assessment, and Thinking Mathematically: Principles for the Design of Model-Eliciting Activities by Richard Lesh, Mark Hoover, and Anthony Kelly in *Developments in School Mathematics Education Around the World*, Vol. 3, National Council of Teachers of Mathematics, Reston, VA, 1993.

Packets Performance Assessment for Middle School Mathematics; Million Dollar Getaway, Educational Testing Service, Princeton, NJ, 1995.

Pursuing Excellence: A Study of U.S. Eighth-Grade Mathematics and Science Teaching, Learning, Curriculum and Achievement in International Context, U.S. Department of Education, National Center for Education Statistics, Washington, D.C., U.S. Government Printing Office, 1996.

Structuring Cooperative Learning: The 1987 Lesson Plan Handbook by D.W. Johnson, R.T. Johnson, and Edythe Johnson Holubec. Edina, MN: Interaction Book Company, 1987.